New Smile and

Better Health in

the 21st Century:

Full Mouth Reconstruction

and Smile Makeovers

Dr. Kathleen Carson, DDS

New Smile and Better Health in the 21st Century:

Full Mouth Reconstruction and Smile Makeovers

Dr. Kathleen Carson, DDS

First Edition
First Printing

evolveALOUD Publishing
3586 Aloma Ave, Suite 12
Winter Park, FL 32792
publishing.evolveALOUD.com
USA

ISBN 978-1-257-06227-0

Printed in the United States of America

Design & Layout:
evolveALOUD Publishing
publishing.evolveALOUD.com

Contents

1

Why I Wrote This Book

You've no doubt heard the expression that "knowledge is power." However you came to be reading a book about Full Mouth Reconstruction - whether a dentist recommended the process or you started researching it on your own – my goal here is to give you both.

Thanks to the internet, you already have access to pages and pages of data covering all kinds of dental procedures and processes. Maybe some of these have been recommended to you, or maybe there are some that you want or think you need. With so much information out there, it can be difficult to know what is valid, not to mention what is going to work best for you.

That's why I wrote this book – to share what is important for you to know that I have learned during my years of studying, practicing and now helping to develop the field of Full Mouth Reconstruction – so that you can take a more active role in determining your dental destiny.

If you're like a lot of people, the idea of getting involved in your health care decisions might be new to you. When we were growing up, we basically gave our dentists and doctors our blind trust. They told us what we needed to have done, and we did it because they told us to.

Today, things are different. Average people are learning more about their health care and using that knowledge to make choices about what is and isn't right for them. At the same time, medical care – and especially dental care – continues to evolve dramatically. With new and different techniques and treatments constantly emerging and improving, you really need to do some research to have an understanding of what options are available that might work for you.

The bottom line is, you only have one smile. If you're going to invest your time and money in improving it, doesn't it make sense to know what you're getting into – and what you can expect to get out of it? That's what I hope to show you over the next few chapters.

So Many Choices…

There's a famous story that was done for Reader's Digest in 1997 about dentistry -- a person went to around 50 different randomly selected dentists, and wound up with a multitude of completely different recommendations for treatment at prices ranging from $500 to $30,000, and everything in between. Back when the study was done, the conclusion was essentially, "Wow, these dentists have *no idea* what they're doing, or they are unethical." The automatic assumption was that the more expensive dentists had "over-diagnosed" the patients.

Today, more knowledgeable dental professionals don't see it that way. It's not that any of the dentists in the study were "wrong," it's just that each of them addressed the patient's needs based on their own understanding of what problems were involved as well as their interpretation of what the patient wanted. Did the patient want to just put out the fires? Did he want to address the causes of the fires? Did he want to prevent future fires? Did he want to improve the look of his smile through elective cosmetic work?

In addition to these very important questions regarding the patient's desires and values, every dentist has their own understanding (which may be limited based on experience and training) about what can or should be recommended in each individual case. Some dentists are trained to simply concentrate on dealing with problems as they occur – filling a cavity here, whitening a stain there. Others have invested in learning about more comprehensive approaches, looking at not only the symptoms, but the causes of dental problems, as well as dental issues that might pop up down the line.

What this all means is that the varying recommendations from the dentists in the Reader's Digest article were based upon all the doctor *and* patient factors above. Today, you will find a very similar scenario.

When a dentist offers a more expensive treatment plan, often they do so because it's the best and most effective plan or it may improve your smile or function in ways you hadn't thought of. It may not only address problems now, but also prevent bigger problems (as well as save you money) later on.

It's up to you to decide how to answer the questions above in regards to how you want to approach your dental health. Let your dentist know what you want, listen to their own recommendations and reasons, and finally decide for yourself what path is the best option for *you*.

My Background

I know that the more labor-intensive option is often the best because I discovered it first-hand. It really goes all the way back to my childhood, and my dad. When I was growing up in Newport Beach, California (not too far from where I live and practice now, in Southern California's Conejo Valley), he used to drive me crazy whenever I would do my homework. I'd finish the assignment and show him. He would look it over, turn to me and ask,

"Is this the best you can do?"

Of course, I'd say "yes," because I wanted to go outside and play sports or hang out with my friends.

My dad, however, wouldn't let it go. He'd usually follow up with something like, "Okay…but once you sign your name to this, you can't take it back. You're telling whomever you give this to that this is the best that you can do. So you'd better be sure that it is."

I'm not going to lie to you. At the time, I didn't exactly appreciate those moments – nor did I have any sort of major epiphany about how important they would be to me later on. I'd sigh and grumble and usually put a little more effort into the project.

However, as I got older, my dad's words did stick with me. Over time, I realized that I didn't feel comfortable putting my name on something that wasn't done right. So I guess you could say he made me strive for perfection.

This desire to do things right also led me to dentistry. I was into health care as a kid – there were a lot of occupational programs offered at my high school where you could learn how to take x-rays or how to be a nurse, and I tried them all. At the same time, I really liked the programs where I could do things with my hands, like home repair, biology lab, and computer technology.

I loved science, so when I did my undergraduate studies at UC Santa Barbara, I majored in Biology. After I graduated (with honors – remember, I strive for perfection!) I searched for a career in healthcare that would give me the opportunity to work with people, work with my hands, do research, teach, *and* have the opportunity to continue enhancing both my education and my profession.

I also wanted a career where I could be my own boss, so that I could control my own professional destiny. In addition to all of this, I needed a "family-friendly" career that would give me the freedom to be involved with my own growing family and their needs.

Dentistry, it turned out, was a terrific fit for me.

Dentistry allows me to be involved in all aspects of care, from research and teaching to designing and building patients' smiles. I enrolled in dental school at UCLA, and, like the over achiever I am, graduated cum laude. I then went on to a general practice residency at the V.A., Sepulveda.

Why FMR?

When I went to dental school, I was most interested in giving people the "Wow" factor –working with my hands to create beautiful, perfect smiles for my patients. I studied cosmetics and aesthetics and learned how to do really good aesthetic dentistry.

However, I found (as lot of dentists did at the time) that just because things looked really good, that didn't mean they functioned well or held up over time. Maybe there was a problem with a patient's jaw joint, maybe they didn't chew quite as well, maybe the profile of the patient could have been improved, or maybe the restorations didn't hold up as well as was expected. Suddenly there were all these other aspects emerging that could be addressed to make the restorations I was doing work better for the patient.

This realization drew me to an area that I never expected to be interested in, cranio-facial functioning. This also meant dealing with pain – which was a surprise because I never really wanted to work with the "pain people," I wanted to be the cool, "she gave me a beautiful smile" person. What I learned was that I had to deal with those kinds of issues in order to create beautiful smiles that functioned well for each individual patient, based on their individual situation.

That meant I needed to start taking more classes. I've really never stopped, and I probably never will be done because we are constantly evolving. I joke with friends all the time, if you don't really want to learn *everything*, don't start down this rabbit hole, because you can't in good conscience provide really good, high-quality dental care without taking the time to really learn and understand how everything works.

Today, I'm a fellow in neuromuscular dentistry at LVI, and right now I'm involved in research that I hope will take the Full Mouth Reconstruction approach to dentistry to the next level, working with a lot of the experts I used to learn from. I still go to a ton of classes, but now I'm involved in teaching too. I am also a member a number of professional dental organizations and societies.

I love that my profession and my career are constantly evolving. However, the best thing about what I do, hands down, is that a lot of times we'll change people's quality of life -- and that is just the coolest thing ever.

We had a patient recently who wasn't sure about restoring her smile. She waited and waited, and she finally decided she wanted to do it -- and it made such a change in her life. She called my office and said that she used to go to her kids' soccer games and sit and not talk to anybody. Now she says she's so much more social, she smiles, she laughs, and she never really realized that the reason she kept to herself before was that she was self-conscious about what she looked like and how she spoke. She told me she's been a member of a lot of social media sites for years, but never, ever posted a profile picture -- until now.

To me, that just means everything.

There are so many stories like that. Full Mouth Reconstruction isn't something little, it's major. If you look at a photo of somebody who's missing teeth, or someone whose smile doesn't look great, and you put that photo next to a picture of the same person after he's had his teeth done, the change is incredible. He looks smarter, he looks more confident, he looks more successful – and he can actually become all of these things. Personally, professionally, financially, the impact really is amazing.

If you undergo Full Mouth Reconstruction, my hope for you is that you will experience the same result. Right now, by reading this book, you're taking the first step in getting involved in your own health care. I want to empower you to take some responsibility for your own conditions and to learn what options are available out there for you and why they might be desirable, as opposed to simply relying on other people to make the decision for you.

I think one of my assistants put it best when she said, "People get their nails done, they buy 25 different purses, they get their hair cut and colored every other month, but the best accessory they walk around with is their smile." You don't get to change your smile every day, or every other month – you do it maybe once or twice in a lifetime. Since your smile is your most important accessory, I can't overemphasize what getting it right means.

So I hope you'll keep reading, and give yourself the knowledge – and the power – to transform your smile, and your life. You deserve the best.

I want to make sure you get it.

2
Why Full Mouth Reconstruction?

If you're not happy with the way your teeth look or function, believe it or not, you're actually pretty lucky. Today, you have more options to change this than ever before. Thanks to advances in modern dentistry, even the most damaged mouth can now be transformed into the smile you've always wanted. As you can probably tell by the internet and local media ads promising "smile makeovers" that can deliver incredible results in a few short visits, cosmetic dentistry is a booming business.

To a person with a less-than-perfect smile, it can all sound incredibly tempting, but there's a problem. In some circumstances, "smile makeovers" can also be dangerous.

Let me explain with an analogy.

Let's say that your kitchen sink is need of repair. The pipes are leaking, the fixtures are worn out and stained, and everything looks terrible.

You could have a plumber come and tighten the pipes to stop the leaking and replace the worn out fixtures with gleaming new ones. Your sink will look as good as new, or even better.

For awhile.

As long as those same old pipes are still there beneath the surface, your beautiful new sink is living on borrowed time. Eventually, it will probably start leaking again. When that happens, the resulting problems may be even worse. You didn't fix the underlying problem that was the actual cause of all the trouble in the first place. Instead, you put your time, effort and money into a faster cosmetic solution that made things look better, but didn't actually fix anything.

"Smile makeovers" can be a lot like that. Yes, there are cases when performed under the right circumstances by a reputable dentist, they can create beautiful, lasting smiles.

However, when underlying problems are present, a quick, cosmetic fix isn't enough.

Those cases call for more comprehensive treatment. They call for full mouth reconstruction, also known as full mouth rehabilitation. Some may simply call this comprehensive dentistry.

Full mouth reconstruction, or FMR, is essentially a series of procedures used in combination to create a healthy mouth. The primary purpose is to deal with "functional" problems that interfere with the health of your teeth, mouth and gums – problems such as gum disease, TMD (temporomandibular dysfunction), teeth and jaws that are out of alignment, as well as other things you may not be able to see, but may be causing problems. It may also involve procedures that are essential to good dental health and also visible, such as replacing missing teeth, replacing failing restorations, fixing broken teeth, repairing wear and tear and moving teeth into alignment.

While FMR isn't purely cosmetic, it can still have major cosmetic benefits. Designing your new smile with your dentist is an important part of the process, and you will likely spend time with your dentist talking about the changes you want to see and your best options for making them happen. The end result is a smile that isn't just more attractive, but healthier as well.

How Does FMR Work?

The term "full mouth reconstruction" can sound like a major undertaking –akin to letting a giant construction crew camp out in your mouth for months on end.

It's true, full mouth reconstruction is not a simple process. It is likely to include several procedures. But that's essentially the point of full mouth reconstruction. In many cases, dental problems don't happen one at a time, or on their own. They often happen gradually and accumulate over time, leaving a patient requiring more than just a simple fix.

Instead of tackling problems haphazardly, or dealing with cosmetic issues and ignoring the underlying problems, FMR deals with the mouth as a whole. An FMR dentist will look at all aspects of how your mouth works and create a blueprint for a healthier, more beautiful smile and the best way to accomplish this goal.

FMR means dealing with the causes of problems instead of just the effects. For example, when a bite is out of alignment, it may lead to jaw pain, uneven tooth wear, or both. Fixing only the unattractive and uneven tooth wear without considering the jaw joint that is out of alignment is the same as making that ugly sink look better without fixing the plumbing problem that caused much of the damage to begin with. While FMR may seem to be expensive or take more time, it is actually more likely to save you time and money in the long term. For example, replacing worn out dental work and dealing with missing teeth while also correcting the way they function together will result in a healthier environment that both holds up and functions much better over time. FMR may involve performing multiple procedures at once because it is the best and most comprehensive way to solve a problem as a whole.

Of course, since there are multiple steps involved in FMR, you may need multiple appointments to complete your treatment. Depending on the treatment necessary to achieve the desired results, it may take

only a few visits, it could take months, or even a year or more before everything is finished. If you don't like to spend a lot of time in the dental office, let your dentist know. With proper planning and coordination, multiple procedures could possibly be done in fewer visits, depending on your individual situation. You may even be able to undergo treatments with sedation if you are one of the millions of people who experiences dental anxiety.

Full Mouth Reconstruction Procedures

Full mouth reconstruction can involve any combination of a wide variety of procedures. Some dentists perform all of the treatment on their own, while others may bring in specialists. Specialists could include periodontists (who concentrate on the gums), orthodontists (who concentrate on moving teeth and possibly broadening dental arches), endodontists (who do only root canal treatments) and oral surgeons (for surgical aspects). It all depends on the specifics of your case and your own dentist's training and specialties.

After an examination of your entire mouth, head, and neck areas, your dentist will formulate specific full mouth reconstruction plans that are possible options for you. Sometimes there may be more than one solution for rehabilitation. For example, if you are missing teeth, the options available to replace those missing teeth may include removable dentures, cemented bridges, or implant restorations. In some cases, there may only be one course of action recommended, or even possible, for your treatment. It all varies on an individual basis.

Your FMR plan may include any combination of the following treatments:

- Teeth cleaning

- Periodontal therapy/treatment of periodontal disease

- New fillings or crowns

- Replacement of old fillings or crowns.

- Repositioning of the jaw

- Oral surgery procedures

- Gum surgery procedures

- Gum tissue contouring

- Temporary restorations

- Permanent restorations

- Orthodontic procedures

- Neuromuscular dentistry

- TMD (temporomandibular disorder) therapy

- Bite "reprogramming" or orthotic therapy

- Dental Implants

- Veneers

- Inlays/Onlays

- Bonding

- Dentures

- Tooth Contouring/Reshaping

Remember, while your dentist is the expert, this is *your* FMR – so be sure to communicate your own concerns and take an active role in designing your new smile.

How the Full Mouth Reconstruction Process Begins

Full mouth reconstruction begins like any other dental procedure – with a visit to your dentist. He or she might recommend a comprehensive rehabilitation to you, or you might bring it up on your own if you know you have multiple issues in your mouth and are ready to tackle them.

Your dentist will next perform a comprehensive evaluation examination, where your mouth will be analyzed and the sources and symptoms of any issues identified so a plan can be formulated to deal with them. This examination will specifically focus on:

- **Your Teeth:** Your dentist will fully assess the condition of your teeth, looking for decay or cavities, wear and tear, cracks, movement, length and spacing.

- **Your Gums:** Your dentist will look for signs of gum disease – symptoms like deep pockets, swelling or bleeding, loss of or buildup of gum tissue and problems with bone density.

- **Your Bite:** Your dentist will examine whether your bite is stable, whether or not it causes you pain and whether or not it is causing unnatural wear and tear on your teeth or jaw joints.

- **Esthetics:** Your dentist will assess the appearance of each individual tooth as well as your smile as a whole in terms of proportion, size, shape and color, as well as how your teeth fit in with your gums, your lips, your mouth, and your face both from the front and from the side.

In order to get a comprehensive picture of how all of these factors work together, your dentist will likely take x-rays and photographs, and take impressions of your teeth to use for models of your teeth and bite. Advanced computerized analysis of how your teeth, jaw joints, and muscles all function together may be necessary. You may be referred to a specialist to deal with any problems that are best treated by an expert – or your dentist may work alone to create a comprehensive treatment plan that will work for you.

Are You a Candidate for Full Mouth Reconstruction?

You may benefit from full mouth reconstruction if you have any of the following:

- Missing teeth

- Cosmetic concerns

- Cracked, broken or injured teeth

- Old dental work that needs replacement

- Wear and tear on teeth due to grinding, clenching, erosion, or poor function

- Jaw pain, muscle pain or headaches resulting from your bite

How Much Does Full Mouth Reconstruction Cost?

Since full mouth reconstruction traditionally involves most or all of your teeth and several office visits, cost can definitely be an issue. Naturally, every case is unique, and costs will vary widely based on what types of treatments are involved in the reconstruction. The overall cost can range anywhere from a few thousand to sixty thousand dollars, and perhaps even more.

The good news is that since full mouth reconstruction is based on procedures that are dentally and medically necessary, there may be ways to find financial assistance. Unfortunately, the idea that dental insurance will help with many of the expenses is a bit of a misconception. It's true that if you have dental insurance, they may

assist you up to their maximum. However, that "maximum" is often only one to two thousand dollars. Of course, this will vary depending on the type of coverage you have as well as the procedures involved with your case. You'll want to check with your insurance provider and your dentist for a clearer picture of what is and is not covered and how much your insurance company will assist you with.

If you do not have a dental insurance plan, you might find it's actually more cost-effective to find other ways to pay for your treatment. Dental insurance can be very expensive, and often has many limitations and may wind up offering minimal assistance. That said, discussing the nuances of dental insurance is beyond the scope of this book and should be something you discuss with your dentist and your insurance carrier.

So what options are available to you beyond dental insurance? Believe it or not, in some rare situations, your medical insurance may assist you with some necessary work. Again, this is beyond the scope of this book, but it never hurts to discuss the possibility with your dentist.

Insurance aside, there are still plenty of other options. Many dentists now offer third party financing where the cost of treatment is broken down into monthly payments that are easier to afford. In addition, many medically and dentally necessary procedures can be considered as tax write-offs. This is something to ask your accountant about. Health Spending Accounts (HAS's) may also offer a good way to pay for treatment.

The bottom line is that if cost is an issue, there may be help available. If your dentist recommends full mouth reconstruction and you're worried about how you will pay for it, be sure to let him or her know. Your dentist will work with you to come up with a plan that works for you.

3
Choosing the Right Dentist for your Full Mouth Reconstruction

Choosing to undergo full mouth reconstruction is a major decision – one that is likely to affect your health, your appearance and even your life for years to come.

That's why it is essential that you choose the right man – or woman – for the job. That means you need to find a dentist who can handle the unique demands of a full mouth reconstruction.

This might not seem especially complicated at first. Most dentists practice at least some of the procedures involved in FMR. If you let your dentist know you're searching for someone to perform a "full mouth reconstruction," he or she may offer to do the work. Of course, if your dentist is the one who suggests FMR, it's highly likely that they'll also suggest that they perform it for you.

If your dentist has the right experience and training in FMR, this could be an ideal situation. However, the mere fact that your dentist offers to do the job doesn't necessarily mean he or she is *right* for the job.

Full mouth reconstruction isn't a new dental *procedure* – in fact, most of the procedures involved have been around for decades. Instead, FMR is a new *approach* to dentistry where the system – mouth, gums, teeth, etc. -- is taken into consideration as a whole, and everything from disease to function to aesthetics, as well as everything in between, is addressed.

A great many dentists, however, still view dentistry as a piecemeal process. Dentists traditionally deal with problems as they pop up rather than first looking at the big picture. That sort of dentist might, for example, offer to treat your gum disease, replace your old crowns and beautify your teeth with veneers, and call that a full mouth reconstruction. Others might perform what they describe as an FMR, but focus only on appearance issues (like quick fix veneers) or only on functional issues (like breaking teeth) instead of addressing how everything works together.

A dentist who understands FMR would approach a dental problem differently. He or she would begin by analyzing both the symptoms and the underlying causes of your dental issue or issues. Then he or she would design a plan to bring all aspects of your dental health into harmony. Instead of putting out fires as they erupt, this sort of dentist will focus on combining the right treatments to create a smile that will function properly, look beautiful and natural, be disease-free, and remain strong and healthy for years to come.

You really don't want to put your dental health in the hands of anyone who will offer you anything less.

Where to Find Dentists Qualified to Perform FMR

Since FMR is a fairly new approach to dentistry, dentists who really know, understand and practice it properly aren't always easy to find. Like a lot of other professions, dentistry is constantly evolving, and advances over the past few years have significantly changed the range of what dentistry can do for you.

There are still plenty of dentists, however, who have been practicing for 20 or 30 years and have no interest in moving beyond "the way things have always been done." These dentists don't learn the new technologies and they don't enhance their education -- a visit to one of them is like stepping back in time to 1982.

In some cases, circa 1982 dentistry may be all you need. However, a dentist like this would not be a good choice for an FMR.

Actually, chances are a dentist like this has never *heard* of FMR.

Of course, just because a dentist keeps up on all the new technologies doesn't mean he or she does good work. The key is to look for someone who has the right combination of training, experience, and results.

You could look into dentists who attended postgraduate institutions that focus on the type of comprehensive reconstructive dentistry involved in FMR. Another good candidate to talk to might be a prosthodontist (a dentist who's a specialist in dealing with missing or problem teeth), who has completed some additional training at a dental school level. The key is an emphasis on the whole, combining function (how things work) and esthetics (how things look).

Your Current Dentist

Your current dentist may be able to recommend a dentist who can perform your FMR, or he or she may meet all the qualifications to do the job. As I mentioned before, if your dentist is the person who recommended FMR in the first place, he or she is likely to expect to perform it, which can be a completely viable solution if your dentist is trained and proficient in FMR techniques. Of course, you should not move forward without first asking the questions we'll get to later in this chapter -- and making sure you're satisfied with the answers.

If you aren't, there is no professional requirement that obligates you to hire your current dentist to perform your FMR. Even if you're inclined to go with that dentist, it's always a good idea to get a second opinion before starting any major medical or dental procedure. Don't worry about offending people when it comes to making sure you get the best treatment available for your smile and your health.

People You Know

Many people find their dentists by word of mouth – asking friends, relatives, coworkers, and even acquaintances about their own experiences. If you know someone who's had FMR and is happy with their treatment, you should definitely ask them who did it, do a little research on your own, and possibly set up a consultation with the dentist.

The Internet

The internet is a great resource for finding all kinds of professionals and provides a good launching pad for your search. Searching under terms such as "full mouth reconstruction" may provide you with websites of dentists in your area.

If there are no qualified dentists in your part of the country, don't panic. Today, many people travel out of their local area for this type of treatment – sometimes even making it part of their vacation plans. If you don't live near the right dentist, finding one in another area may be your best option. Remember, your health should be your first priority, not your commute time.

Researching your FMR Dentist

When preparing for comprehensive dental treatment like FMR, it's wise to get as much information about prospective dentists as you can. You can learn a lot about the dentist, the team, and the office even before you set up your initial consultation appointment. A good place to start is online -- most dentists have websites that tell you about their training, their team, their office, and what to expect.

Just be aware that these websites are designed to offer more than information – a well-designed website is also a marketing tool. That doesn't mean you shouldn't use the information provided on a dentist's website; you absolutely should check out the website of any dentist you find through other means as it likely contains many of the answers you need. However, reading through a dentist's carefully crafted web content should be the *beginning* of your research, not the end of it.

If you're researching a dentist who doesn't have a website, call the office and ask the receptionist for the information you need. If she cannot provide it, she should be able to connect you with someone who can, or take a message and get back to you within a reasonable amount of time. If you don't get an answer, cross that dentist off your list. You want to be able to trust your smile to someone who has an effective, well-trained team – another point we'll get to a little later.

The Right Qualifications for FMR

Since full mouth reconstruction is complicated and involved, you want a dentist who has experience in this area. You definitely don't want to be a dentist's first-ever FMR, or even, if at all possible, their 10th FMR.

Keep in mind that more years in practice doesn't necessarily correlate with more FMR experience. As I mentioned before, there are many professionals, across the board, who don't continue their education beyond basic training. Conversely, a person could be in practice for five years and spend more time in continuing education and specialized training than the first person would in a lifetime. The goal is to find a combination of the best, state-of-the art training along with experience in practice. Don't be satisfied with one without the other.

The right training is crucial when choosing an FMR dentist. Look for a dentist with an advanced postgraduate (after dental school) education at a reputable institute that requires many hours of hands-on training time. The achievement of a fellowship or accreditation at this postgraduate level indicates that a dentist has passed certain testing criteria. Additionally, this extensive training shouldn't stop with your dentist. The right dentist for FMR will be surrounded by a well-trained team that is also experienced in full mouth reconstruction procedures.

This type of training is time-consuming and costly, which leads me to an area that should *not* be a priority when researching FMR dentists – price. You've probably heard the saying "you get what you pay for," and full mouth reconstruction is not an area where you want to get less! You want your FMR to be done right the first time so that it looks beautiful, functions properly and holds up for a long time. You don't want to skimp on the materials, the labs or the training of the people performing it. Price shopping is not a good way to get the results you need and deserve.

No one else's FMR will be exactly like *your* FMR. The best way to get an idea of a prospective dentist's work is to see before and after pictures of patients. Not only will you get a clearer picture of how a dentist works with each patient's teeth, gums, facial structure and more, you'll also get at least an idea of the range and scope of the work he or she has done. Don't just settle for photos on a website – ask to see actual photos of actual patients. You can also ask a prospective dentist to share some of the stories behind the photos to learn how he or she approached other cases.

Since FMR is modern, state-of-the-art dentistry, you should expect any dentist performing it to be up-to-date on the latest technology and techniques. This can include sedation and relaxation options for those patients who may be nervous or uncomfortable – if you are one of them, don't be shy about asking how your comfort will be addressed during your treatments.

Some practices also offer extra touches ranging from video and music for your entertainment to massaging chairs. The absence of these bells and whistles shouldn't be a deal-breaker if you've found a great dentist; however, it's nice to know exactly what will be available to you when you put your smile in his or her hands.

You shouldn't worry about asking your potential dentist to supply you with the information you need to make a decision. You want to know about the person and the procedures that could potentially alter your life (in a good way!).

Questions To Ask a Prospective FMR Dentist

Here is a list of potential questions to ask a prospective dentist before going forward with treatment.

1. Where did you get your training?

2. What type of advanced training do you have?

3. Has your team been trained?

4. What is the process of getting a second opinion or becoming a new patient?

5. Can I see some before and after patient photos?

6. Can I speak with any of your former patients?

7. What will you do to make sure I am comfortable during and after my procedures?

8. Do you handle all procedures yourself, or do you work with other doctors or dentists?

These are only questions to ask a *prospective* dentist, so they are fairly general. If you consult with a dentist about your FMR, you will obviously have other questions that are more specific to your individual treatment.

The goal is to find the dentist whom you feel most comfortable and secure with, and who has the right combination of extensive training, hands-on experience and positive results. You may need to consult with different dentists before you make a decision. The more you know about a potential dentist and the procedures being recommended, the better your chance of having a pleasant experience with a great outcome.

Isn't that what you and your smile deserve?

4

Getting Started

Once you've done your research and found a dentist (or dentists) you're interested in meeting with, the next step is to make an appointment to get the FMR process started.

Ideally, your first appointments should be the final steps in the process of choosing the dentist who will design and perform your FMR. You'll want to have the clearest picture possible of the people who will be responsible for transforming your dental health, as well as a general idea of how your case will be handled. So, if you can, you should arrange a visit at each office you're interested in to learn more about how they do things before you make your selection.

Some dentists' websites have forms you can fill out to begin the process of becoming a patient, often offering some sort of consultation. In most cases, however, I think it's better to call and talk to a trained team member; that way, you can explain what you are looking for and what you suspect you need.

Don't worry about going in with a specific "diagnosis" of your dental issues, just a general idea of what's wrong and what you're hoping to improve. Ask what types of introductory appointments are available for a patient considering FMR. There are dentists who offer a free consultation appointment, but others do not – and that fact alone definitely should not cause you to turn away. Remember, you need to be looking for the best dentist for the job, not the best "deal."

Some offices offer what I call a "meet and greet" appointment, which is essentially a short meeting, often at no charge, where you can visit a dental office and get answers to your initial questions.

These meetings tend to cover topics like how an office works, the options they offer, policies and procedures, and more. They are typically handled by a key team member rather than the dentist. This type of appointment can give you a feel for what your experience at an office will be like in terms of atmosphere, services offered and overall philosophy and approach. If you are comfortable with what you hear and see at a meet and greet, you'll probably want to set up an appointment with the dentist.

Sometimes a quick meeting with the dentist is possible at this point at either no or a very low charge. This is a great chance to find out if you feel comfortable with a potential dentist before you go through an entire consultation. Do remember that dentists have busy schedules, and ask to book this time in advance when you set up your initial meet and greet.

Once you've had a consultation with a dentist where you've discussed your case and received some specific treatment recommendations, you can also book a "second opinion" appointment with another dentist. This is a good way to find out if another dentist generally agrees with your diagnosis and recommendations. You can get an idea of whether or not you're on the right track and considering the right procedures. This is also a good way to meet and compare other dentists.

You may be charged for this type of appointment, or it may be done as a "free consultation" to give another office an opportunity to meet you and interest you in their services.

The Comprehensive Examination

After you've chosen a dentist to trust with your FMR, the next step before treatment is a comprehensive examination. You might compare a comprehensive exam to a complete physical at your doctor's office; it's more than a simple "checkup." It is a thorough examination designed to provide a detailed view of your masticatory (chewing) system, how it functions, and pinpoint any problem areas.

Here is a basic outline of what to expect from a comprehensive exam:

Photos: Your dentist will need a complete photographic record of your mouth as it is right now, to use in diagnosing conditions and preparing for treatment (and also for the eventual "before and after" photos to illustrate the changes in your smile). You will likely have both extra-oral (outside the mouth) and intra-oral (inside the mouth) photos taken from several different angles to document your current situation.

Extra-oral photos are essentially what you probably expect – just photographs of your face and mouth taken from different angles. You may be asked to smile in some photos and keep your mouth closed in others for a complete view of how your mouth looks from the outside.

Intra-oral photos are taken with a special intra-oral camera that takes a "video tour" of your mouth as well as still photographs of your teeth and tissues. These will allow you to see exactly what the dentist sees when he or she looks inside your mouth and better understand any conditions you might have.

X-rays: You will also likely have a full set of what dentists call radiographs, and you likely call x-rays, taken. X-rays are done to show conditions beneath your gum line, underneath the surface of your teeth, and in between your teeth. They are a very important part of the diagnostic process and a proper exam cannot be done without them. In our office, we use digital film for our x-rays, which reduces a patient's exposure to up to 80% less radiation than traditional x-rays. If you're concerned about radiation exposure, ask prospective dentists if they offer digital x-rays. Digital x-rays also don't require processing or development like standard x-rays, so there's no wait time to see your results.

Impressions: Your dentist will have an assistant take impressions of your teeth and gums to form a model of your mouth to study and work from.

If you've never had impressions taken before, the process is fairly straightforward. Horseshoe shaped trays designed to fit comfortably over your teeth and gums will be placed over your teeth. These trays will be filled with a liquid material that solidifies in a few minutes, forming an "impression" of your teeth. The trays will then be removed and the impressions will be filled with a plaster-like substance that will harden and form a model of your teeth.

Complete health history: You may not realize it, but the health of your mouth and the health of the rest of your body are closely related. Many diseases and conditions you might typically associate with the rest of your body can also affect your dental health, including:

- Kidney disease
- Diabetes
- Sleep Disordered Breathing (Sleep Apnea)
- Artificial valves and joints
- Neuromuscular diseases and conditions
- Cardiovascular disease and conditions
- Respiratory disease
- Immune disorders
- HIV and AIDS
- Gastrointestinal disorders
- Blood disorders
- Sexually transmitted diseases
- Head and neck cancer
- Vitamin deficiencies
- Pregnancy
- Bulimia
- Tobacco use
- Bisphosphonate treatment

This is only a small list of potential conditions that your dentist may need to be aware of. It's crucial that your dentist is aware of conditions like these – or any others – that may be contributing to your dental problems or play a role in your treatment options.

At the same time, research has shown a link between dental problems – particularly gum disease – and health problems in the rest of your body. These include:

- Heart disease
- Stroke
- Kidney disease
- Premature birth
- Diabetic complications
- Respiratory disease

In other words, a complete health history is not only necessary, it's also very relevant.

Complete dental history: As you might expect for a comprehensive dental treatment like FMR, your dentist will need to know about any past conditions or current conditions. If you can obtain any recent x-rays from other dentists, they can be helpful. Your perspective is also an important part of the examination. Your dentist will likely ask questions including:

- Are you having any discomfort?
- When do you feel this discomfort?
- Do you know what causes your discomfort?
- How long does the discomfort last?
- What does your discomfort feel like?
- Are you able to chew and function comfortably?
- Is there anything else about how your teeth or gums are functioning that you don't feel comfortable with?

Since symptoms of problems may show up as pain or discomfort, how you are feeling is an important tool during the diagnostic process. The absence of symptoms, however, does not mean there are not problems present. Often times, dental disease can be pain free.

Gum Health Analysis: The official name for your gum tissue is "gingiva;" the supporting gum tissue and bone around your teeth is collectively referred to as "periodontium". Any presence of disease of your gum tissue or supporting bone is called "periodontal disease." Inflammation of your gum tissue is called "gingivitis".

Your comprehensive examination will include a complete periodontal exam that will be performed either by the dentist or the hygienist working with the dentist. Whoever performs the examination will evaluate the health of your gum tissue and supporting bone by taking measurements of your current bone level, gum tissue "pockets" and recession areas (places where gum tissue and bone levels have been lost around teeth). Any mobility of your teeth will be assessed, along with the occurrence of any bleeding of your gum tissue, the amount of plaque and calculus (hardened plaque, also called tartar) around your teeth, and the presence of harmful bacteria.

This exam will provide a full picture of the current health of your gum tissue and supporting bone (your periodontium).

TMJ analysis: TMJ is short for temporomandibular joint – the joint that connects your skull with your mandible, or jawbone. Your jaw joint is a ball and socket joint with a cartilaginous disc between the ball and socket that is held together by muscles and ligaments. If your jaw is "out of alignment" due to a problem with one or both of the joints, this is similar to having a car axle being out of alignment – much more damage will result from this misalignment over time.

A TMJ analysis will determine if your jaw joints are working properly. Your dentist will assess your range of motion and look for symptoms of TMJ dysfunction, including clicking and popping, deviation, locking in an open or closed position, crepitus (grating) sounds and any pain or dysfunction. This part of the examination will also include palpation of the muscles associated with your jaw joints so that any possible problems with the muscles can also be assessed.

Bite analysis: This is generally what it sounds like – an examination of how your teeth fit together. The term dentists use for this is "occlusion," so if you hear the word, your bite is being discussed. Your bite will be analyzed in several areas including:

- How your teeth are positioned in relation to each other – i.e. crowding or spacing
- How teeth come together to chew and function
- How teeth overlap
- Patterns of wear or fracturing caused by the bite being off track
- How the bite functions with the position of the jaw joints and the muscles that hold everything together

Bite analysis goes hand in hand with TMJ analysis – the ideal is for your teeth, jaw joints and muscles to work together in harmony.

Sleep apnea analysis: If you suffer from sleep apnea (periods where you stop breathing during sleep), this can adversely affect your FMR, not to mention your overall health. Be sure to tell your dentist if you suspect you may have sleep apnea – if he or she suspects it, you may be asked a few questions about your sleep during your appointment.

Sleep apnea can often be successfully treated with an oral appliance, so dealing with the condition – if you have it – will likely be a part of your FMR.

Analysis of condition of teeth: Each individual tooth in your mouth will be looked at and assessed as its own entity in the following areas:

- What is the current condition of the teeth?
- Are there any restorations present (i.e. fillings, crowns, etc)
- How are the restorations holding up?
- Is there any decay (cavities) present?
- Is there any fracturing?
- Is there any wear on the teeth?
- What is the general health of the tooth?
- Are any teeth missing, and if so, has this affected other teeth?
- Are there any implants or replacement teeth (dentures or bridges) present?

Each of these elements is looked at individually and as a whole – that way your dentist can see each piece of the puzzle while also analyzing the whole picture.

Cosmetic Analysis: If there is anything you don't like about your smile, this is where those issues will be addressed. During the cosmetic analysis, you can tell your dentist what you'd like to change about your smile. If you aren't sure what you don't like, don't worry. The right dentist will know a lot more about what makes a smile attractive – or unattractive – than you do. He or she will also know how to work with your unique teeth, jaw and facial structure to create a beautiful smile that suits you.

The cosmetic analysis is where your dentist will analyze the general appearance of your smile and focus on specifics including the color of your teeth, the position of your teeth, whether any teeth are worn or fractured, the shape of your teeth both individually and as a whole, and the length and width of each individual tooth and how they fit together to create your smile.

Many top cosmetic dentists have received training in what is called "the golden proportion" (also called the golden ratio) – a mathematical formula used in everything from mathematics to art for determining the right balance of elements to achieve the most aesthetic and pleasing proportions. This formula provides dentists with a starting point to designing the perfect smile for you.

During the cosmetic analysis, your dentist may also perform a phonetic (how you speak) analysis, looking at how the shape and position of your teeth affects your speech. Specifically, your dentist will look at where your teeth meet your lips to make sounds like "f", where your tongue touches the back of the teeth to make sounds like "t" and how the edges of the upper and lower teeth come together to make "s" sounds.

Final Overview: The last part of your comprehensive examination should be a wrap-up of what to expect from your treatment. You should be fully aware of:

- what is happening
- what procedures will be done and when
- what the outcome is predicted to be
- how long it will take
- how much it will cost and when that payment is expected

You should also make sure you discuss any other important details, including whether or not you will be using sedation and the guidelines for that, or if will you need to see a specialist and when. Make sure you also talk about any dental insurance assistance you might have and other potential payment options.

This final wrap-up may be delayed, because in some cases it is necessary for the dental team to take all of the data that has been gathered during the appointment, sit down and study it. There may be more than one way to treat your conditions – if so, your dentist will discuss various options with you to help you achieve the right results in terms of health and function as well as aesthetics.

This is another time to remember that in dentistry, you "get what you pay for." Your dentist may present cheaper options that don't function or look as good as alternatives – be sure you have all the relevant information before you make a decision based on price alone. Something that may cost less now may require replacement (at additional cost) sooner, may not function as well, or may not look as good.

After the dentist has had enough time to analyze all the pieces of your puzzle, you can expect a follow-up appointment where you will go over his or her findings as well as recommendations for treatment, any alternatives, and what you can expect from each potential solution.

It is then time to make some decisions and begin moving forward in creating your new, healthy, functional, beautiful smile.

5
Gum Problems

A Full Mouth Reconstruction may be all about your teeth, but the first and possibly most important step in the process is securing the health of your gums – as well as the bone that supports your teeth. Healthy gums are the foundation of good oral health, because together with the supporting bone, they hold your teeth in place. When they are infected with periodontal (gum) diseases like gingivitis and periodontitis, it eats away at that support system and can ultimately lead to tooth loss if the condition isn't treated. In fact, periodontal disease causes more tooth loss than cavities in adult patients.

Periodontal disease is an infectious disease caused by harmful bacteria in your mouth that can spread from your gums to the supporting bone. It can enter your bloodstream and affect other systemic health conditions throughout your body, and can even be passed on from person to person through direct contact with saliva.

When gum disease causes inflammation of your gum tissue only, it is called gingivitis. This is the early phase of gum disease that often leads to the more advanced condition called periodontitis. Periodontitis is when the infection has spread from just the gum tissue into the supporting bone of the teeth as well, resulting in bone loss around the teeth. Periodontal disease can be generalized (around all of your teeth) or localized (around only some of the teeth). It ranges from mild (only a small amount of bone has been affected) to moderate to advanced (at risk of losing teeth). How quickly periodontal disease progresses from mild to advanced is varies greatly, depending on the host conditions (that's your mouth and body) that it is thriving in.

More than 85 percent of people have some level of periodontal or gum disease, and since the gums are an essential component of your oral health, this can be a major problem. The good news is, thanks to advanced training and technology within the dental profession, gum problems can now be diagnosed earlier and treated sooner. If your gum disease has advanced beyond the early phases prior to being detected by a dental professional, there are treatment options that can help stop its progression and maintain your health. It's important to know that periodontal disease is not cure-able; once you have it, you will always be susceptible to it. It is, however, controllable with proper professional and home care.

Periodontal infection is caused by bacteria that are always present in our mouths; it can be kept in check through regular maintenance. Infection occurs when those bacteria and the toxins (poisons) they produce reach the point where they overwhelm the immune system. The lining of the gum tissue next to the tooth then becomes cracked and ulcerated and can bleed when touched. Left unchecked, the toxins move into the root surface of the tooth, and if the infection is allowed to continue, begin to destroy the bone that supports the teeth.

The bone loss associated with periodontal disease may ultimately result in the loss of teeth since there will be no bone left to hold them in place. This can happen when periodontal disease isn't caught at a mild or moderate phase, treated, and controlled with proper maintenance. With regular maintenance and proper home care, periodontal infection can be controlled.

Some people are more susceptible to periodontal disease than others, depending on how strong their immune system is and how well they fight infection. The tricky part is that everyone's ability to fight off infection varies from day to day and week to week depending on what else is going on with their lives and their health. These normal variations mean that even people who generally take good care of themselves and their teeth can develop periodontal disease.

What Causes Periodontal Disease?

The main cause of periodontal disease is bacteria. Initially, these bacteria form a plaque, a sticky, colorless film that constantly forms on your teeth. This plaque, if not removed, will harden into what we call calculus or tartar. The harmful bacteria reside in this hardened structure, out of your reach. Regular brushing, flossing, and other home care must be paired with regular dental visits to help to assure that this plaque doesn't build up, harden, and harbor the disease causing bacteria. Even the most rigorous at-home regimen can't always get at plaque below the gum line and between teeth; it can really build up if you miss out on professional cleanings at your dental office.

Every person is not the same when it comes to how often they need a professional cleaning. The old "see your dentist twice a year" philosophy is not practical because no two people are alike -- each individual patient needs to be seen and treated in the way that is best suited to their individual health. A dentist or hygienist is the only person qualified to determine how often you should have your teeth professionally cleaned, and this schedule will likely vary as well as conditions in your body are constantly changing. A good dental team will constantly monitor the ups and downs of your dental health and provide the treatment you need to keep it healthy.

Regardless of how hard you and your dentist work to maintain that health, even the most conscientious patient can experience gum disease due to other factors that can also affect the strength of your immune system and the health of your gums. These include:

- **Smoking or other tobacco use**
  Not only is tobacco linked to a laundry list of serious illnesses including cancer, heart disease and lung disease, recent studies have shown that tobacco users have a dramatically increased risk of developing gum disease.

- **Genetics**
  If you have problem gums, you might be able to blame your parents for them. Studies have shown that close to 30% of people are genetically predisposed to developing gum disease, and those people could actually be six times more likely to get it than people without a family history of the problem. There is now a test to determine if you carry the gene that makes you more susceptible to periodontal disease – if you know you have it, you can step up your maintenance routine to help ward off the disease prior to developing it! Ask your dental team if this is something you are interested in finding out about.
- **Puberty, Pregnancy and Menopause in Women**
  If you're a woman, hormonal changes can wreak havoc on your gum health regardless of how well you take care of yourself.
- **Stress**
  Research has shown that stress takes a toll on your immune system, which can make it harder for your body to fight infections like periodontal disease.
- **Medications**
  Birth control pills, some heart medications, anti-depressants and other drugs can affect the health of your gums.
- **Teeth Clenching or Grinding**
  Grinding your teeth at night or clenching under stress puts excess force on the periodontal tissues that support your teeth.
- **Diabetes**
  Diabetics have a higher risk of developing infections – which makes them more susceptible to periodontal disease. Studies have shown that not only are diabetics more likely to have periodontal disease, but also that diabetics with periodontal disease are more likely to have a hard time controlling their diabetic conditions.
- **Poor Nutrition and Obesity**
  A diet that's low on nutritious foods deprives the body of the vitamins, minerals and proteins it needs to stay strong.

This can also make it harder to fight off infections like periodontal disease.

The Mouth-Body Connection

Periodontal disease can affect your gums as well as the rest of your body. It's been shown that the harmful bacteria in your mouth, through periodontal disease, can gain access to your blood stream and airway, affecting other parts of your body and systemic health. This is what dentists refer to as the "mouth-body connection," and it actually works both ways. Left unchecked, the bacteria that cause periodontal disease can affect the rest of your body in serious ways, and have been linked to cardiovascular disease, stroke, Alzheimer's disease, Type II diabetes, respiratory diseases, and prematurity and low birth weight in babies.

When we treat periodontal disease, we're really looking out for your overall health, not just your teeth and gums.

Diagnosing Periodontal Disease

You may have periodontal disease and not even know it; gum disease can occur in your mouth without leaving behind any telltale symptoms. It can progress slowly, without pain. Some people do experience the bleeding gums or soreness and tenderness that have proven to be clear signs of gum disease, but others do not. Without pain or bleeding to alert them to a problem, the average person won't know they have one, leaving the harmful bacteria to continue to multiply.

There are a few symptoms that can be a clear indication of gum disease.

Those include:

- Bleeding during brushing or flossing your teeth

- Swollen or tender gums

- Persistent, unexplained bad breath (also called halitosis)

- Teeth that feel loose

If you are experiencing any of these symptoms, make an appointment with your dentist to have them checked out as soon as possible.

Fortunately, even without symptoms, periodontal infection can now be detected and treated early. When you see the Hygienist as part of your Full Mouth reconstruction treatment, the examination should include a periodontal screening to detect early signs of this infection. If your gums bleed during this screening, that indicates an infection is present. If you catch it early, it can be treated simply and without surgery, and you should be able to move on to the next steps in your FMR fairly quickly. However, if the infection has progressed, surgical intervention may be necessary to control it – and it will almost certainly need to be controlled before you progress with your FMR.

A Word about Hygienists

Just like dentists, not all hygienists are equal. They too need to undergo continuing education courses to keep their licenses in good standing. Like dentists, some may just work to meet the basic requirement so that they can keep practicing, while others will use the requirement for continuing education as an opportunity to stay on top of the latest techniques and technologies.

Top hygienists – the kind that are essential to a successful FMR team – seek out courses on everything from periodontal disease and treatments to the newest restorative aspects of dentistry (and how to best maintain them!). Clearly, hygienists like these offer much more in terms of training and ability than many more "traditional" hygienists.

As a general rule, however, more advanced dental offices will have more advanced hygienists on staff. More advanced doctors expect and demand more from their hygiene team, and at the same time, more advanced hygienists most likely will not be happy working for dentists who are not at the top of their profession.

Treating Periodontal Disease

Your dental office may treat your periodontal disease, or you may be referred to a periodontist. A periodontist has completed three years of additional training beyond dental school with an emphasis – as you might expect – on periodontal disease. Very often, more advanced dental offices (with a well trained hygiene department) can successfully treat and control your periodontal disease, especially if it's not yet in the advanced stages and can be done with non-surgical techniques. If a surgical procedure is being considered or your case is more advanced, you likely will be referred to a periodontist.

The goal in treating periodontal disease is to reduce the amount of bacteria and poisons back to a point where your immune system can take over and keep them under control. This will restore the affected area to health, and can be accomplished non-surgically in many cases, especially when the infection is caught early. Dentists now have access to new and effective technologies that make the treatment of periodontal infection both more complete and more comfortable at the same time, so there's no need to worry about pain. Remember, if you're a sensitive or nervous patient; be sure to discuss sedation options with your dentist.

Treatments include:

Non-Surgical Treatments, also known as Periodontal Therapy

As every individual and their case is unique, the best approach to treat your periodontal disease will need to be determined by your dental team. Many times, a combination of procedures is used to combat the disease.

- **Scaling and Root Planing**
  This describes the use of specialized instruments to scrape plaque and calculus (tartar) from and tooth root surfaces, then smoothing the tooth roots where plaque and tartar accumulate. An ultrasonic scaler, which makes the process more comfortable and effective, may be used with or without other instruments. Following scaling and planning, most offices irrigate the pockets between the teeth and gums with antibacterial or antimicrobial solutions to wash away debris and sterilize the area.
- **Medications**
  Your dentist or hygienist may apply a prescription antimicrobial medicine or a localized, time-released antibiotic directly to your gums to help kill more of the bacteria that cause periodontal disease. Your dentist may also prescribe a rinse or solution to use at home to keep fighting your infection.
- **Laser Therapy**
  Many modern dental offices are discovering that using high-tech lasers in treating periodontal disease makes that treatment much more comfortable and effective. A soft tissue laser will disinfect the gum tissue pocket and seal blood vessels, nerve endings, and lymphatics so that there is no infection, bleeding, or swelling following laser treatment (officially known as sulcular debridement).

Many patients don't require surgical procedures once the above treatments have been completed – besides ongoing maintenance of course. In order to make sure your periodontal infection remains under control, you need to remove bacteria at home every day. This is absolutely critical! Your dentist and his or her staff will formulate a plan for you to follow at home.

There are cases, however, where more invasive treatment is required.

Periodontal Surgery

The first step in fighting periodontal disease, even if surgery is indicated, is usually the above non-surgical periodontal therapy options. These are also done prior to many surgeries to ensure the best outcome possible for the surgery.

If the infected tissue in around your teeth can't be fixed through the above non-surgical periodontal therapy techniques, surgery may be the next step. Your FMR dentist may elect to bring in a periodontal surgeon to handle this aspect of your treatment, and if that is the case, make sure you feel comfortable with the specialist as well.

Periodontal Surgery will likely fall into one of four categories:

- **Pocket Reduction Procedures**
  Gum tissue is folded back so that bacteria and damaged bone can be removed. The tissue is then secured back in place.
- **Regenerative Procedures**
  Membranes, bone grafts and proteins that stimulate tissue growth are used to regrow healthy bone and gum tissue.
- **Crown Lengthening**
  Removing excess gum tissue and/or bone to expose more of a tooth. This can also be used to correct a "gummy smile."

- **Soft Tissue Grafts**
  Used to cover roots or grow new gum tissue where tissue has receded.

Keeping your Gums Healthy

It's important to recognize that periodontal disease is not curable. The bacteria involved occur naturally in everyone's mouth, so the infection can always return, especially if you are susceptible. The good news is that it is controllable. Once your in-office treatment has been completed, your dentist and his or her team will provide you with a plan to keep your gums healthy and keep periodontal disease from returning. Often this means you will need to be seen for maintenance visits on a more regular basis, with treatment being provided at those visits that is above and beyond the type of care someone without a history of periodontal disease would get.

Research has shown that it takes around 45-90 days for the bacteria that cause periodontal infection to produce disease. You'll need to be especially vigilant at home, between your professional maintenance visits, to be sure your infection does not come back. That means practicing good oral hygiene, brushing at least two or three times a day with a soft-bristled toothbrush. Daily flossing, antibiotic or

antimicrobial treatments, an oral irrigator to continue to wash excess bacteria out of your mouth, and a variety of other home care devices may be recommended for you to use. Depending on your individual needs, your professional dental team will recommend what types of home care devices are best used in maintaining your conditions. Your dental team will also see you for more frequent maintenance visits to monitor, spot treat, and help you control the periodontal disease and maintain your health.

Follow the instructions of your hygiene team to keep your gums disease free. Your mouth is now ready for the next step in your FMR.

6
Neuromuscular Dentistry

If you're looking into a full mouth reconstruction, one of the most important things to consider, both at the beginning of your treatment and as an ultimate goal, is how your jaw system functions as a whole.

You can undergo all the procedures you need to get rid of decay, fix broken and problem teeth, replace missing teeth, treat gum disease and make your smile look great -- but if your joints, teeth, and muscles don't work together properly, chances are your FMR won't either. The work done may not last as long as it could, or you could experience pain or dysfunction. That's why a *real* full mouth reconstruction doesn't just focus on your teeth and gums, but your bite as well.

There is more than one way to approach how to rebuild a bite, and no one single way to do things for every patient. The only way to figure out exactly what should be done for *you* is to analyze *your* bite in detail. This is where neuromuscular dentistry can be important.

You may have heard of neuromuscular dentistry as a way to treat TMD – also known as TMJ – a painful condition causing a wide variety of unpleasant symptoms that may not even seem to be related to dental health.

Some of those symptoms include:

- Jaw pain
- Headaches
- Facial pain
- Pain in the back, neck or shoulders
- Numbness in fingers or arms

- Popping or clicking of the jaw
- Teeth grinding or clenching
- Ringing or congestion in the ears

If you have any of those symptoms now, neuromuscular dentistry is likely to be an important part of your FMR. However, even if you don't suffer from TMD or any of the above symptoms, a neuromuscular approach to your FMR can be very valuable.

Looking at the way your jaw joints, muscles and tissues and the position of your teeth function as a whole can insure the best results possible in a complicated FMR. After all, if you're getting the dentistry done anyway, shouldn't the goal be for the results to look, work and feel as good as they possibly can?

What is Neuromuscular Dentistry?

The main difference between neuromuscular dentistry and traditional dentistry is focus – regular dentists work primarily with the teeth and gums, while neuromuscular dentists also consider your nerves, joints, muscles and how your upper and lower jaws are aligned.

When you're at rest, the multiple muscle groups on each side of your face and neck that hold your jaw in place and allow it to function are relaxed. Those same muscles contract when you chew, swallow or bring your teeth together. Ideally, the muscles contract evenly, and your teeth should touch evenly. When your teeth aren't properly aligned so they don't touch evenly, this forces your muscles to shift into an unnatural position to bring your teeth together.

The muscles of the face and neck are often "programmed" to control head and jaw posture in a way that accommodates how the teeth fit together, even though that position may not be ideal. The Neuromuscular Dentist wants to relax these often tense muscles to find their true resting state and establish the bite at that position.

Chances are you don't notice this unnatural movement, even though it forces your muscles to work harder and may be causing unnatural wear on your teeth, strain in your jaw joint, or pain in your muscles. You might notice those symptoms of TMD mentioned earlier, or you might not notice anything at all.

Misalignment can lead to problems you don't see or feel, but that can affect everything from your appearance to your health. The goal of neuromuscular dentistry is to first identify any problems you might be having with your bite, muscles or joints and then design a plan to correct them.

To do this, neuromuscular dentists apply the basic laws of physiology and anatomy to the jaw – measuring the current position of the jaw, called the "physiological rest" position, and determining the ideal position of the upper and lower jaws. Modern technology can help neuromuscular dentists make these determinations, with many of them using computerized instruments to get the most accurate measurements of your jaw's position and movements possible, as well as to treat some problems.

These technologies include:

- **Electromyography (EMG)**
  Used to measure muscle activity in the jaw.
- **Computerized Mandibular Scan**
  Used to precisely measure a patient's jaw movements as well as the resting position of the jaw. A Jaw Motion Analysis (JMA) may also be performed.
- **Sonography**
  Used to record the sounds the jaw makes while moving, including clicking, popping, scraping and grinding, which would indicate a problem with alignment.
- **Joint Vibration Analysis (JVA)**
  Used to identify problems in the function of the TMJ joints.
- **Ultra low frequency transcutaneous electrical neuromuscular stimulation (TENS)**
  Delivers a mild electrical stimulus to the muscles via neural

pathways, inducing involuntary contractions of the facial and masticatory muscles. TENS is used to help overcome the "programmed" muscle memory of the head and neck muscles. A secondary use of low frequency stimulation is to achieve drug-free pain relief of pain of muscular origin.

Neuromuscular Dentistry Treatments

After your bite has been analyzed, your neuromuscular dentist may need to design a treatment plan to bring your bite back into alignment and relieve any pain or symptoms.

If part of your full mouth reconstruction plan involves repositioning your bite, the use of an appliance such as an orthotic, bite plate, or splint will likely be necessary. A bite plate, splint, or removable orthotic is a piece of acrylic designed to bite against. It places your jaw in the more desirable position only while it is being worn. A fixed orthotic can be bonded directly to the teeth to work constantly to reposition the jaws. Your neuromuscular dentist will determine which appliance is right for your treatment.

Braces or other orthodontic treatments may be used to reposition your bite as well. They can move teeth into proper position to change your bite and take pressure off your jaw joints and muscles.

Whatever treatment program your dentist determines is right for you, neuromuscular dentistry will restore the systems in your mouth to functionality and create the best environment possible for whatever other treatments are part of your FMR.

7
Sleep Apnea

You might be surprised to see a chapter devoted to sleep apnea – something most people regard as a medical problem – in a book about advanced dentistry procedures. In fact, there are two reasons why sleep apnea can be a major factor in Full Mouth Reconstruction and needs to be addressed here.

The first reason is that sleep apnea can affect your bite, the wear and tear on your teeth and even your jaw joints. This is probably why the majority of patients with TMD problems also suffer from sleep apnea. In addition, people with sleep apnea are six times more likely to clench and grind their teeth.

If you undergo Full Mouth Reconstruction without dealing with your sleep apnea issues, once the FMR is finished you are much more likely to do damage to whatever work has been done, and may need to replace them sooner. It is also much more likely that you can have functional problems with your new restorations from the beginning.

The second reason sleep apnea should be considered as part of an FMR is a more positive one: many of the most effective treatments for sleep apnea are actually performed by dentists. This means that if you have sleep apnea and are also undergoing Full Mouth Reconstruction, your dentist can likely make helping your apnea problem a part of your treatment plan.

What is Sleep Apnea?

Sleep apnea is a condition that causes a person to stop breathing for 10 to 30 seconds at frequent intervals during the night. People who suffer from sleep apnea can stop breathing like this hundreds of times every night while they sleep, depriving their bodies of oxygen every single time. It's a very common problem, affecting more than 12 million Americans according to the National Institutes of Health. However, since apnea occurs while its victims are asleep, many people don't even know they have it. This can be extremely dangerous since this consistent deprivation of oxygen can lead to cumulative, long-term effects including diabetes, high blood pressure, kidney failure, stroke and heart failure. In other words, sleep apnea can be deadly.

Sleep apnea can affect anyone from children to seniors; however, those most likely to suffer from it are male, over 40 and overweight. It can be caused by a variety of factors, including:

- Allergies

- Large tonsils and/or adenoids

- Genetics

- Enlarged tongue

- Poor Posture (either while awake or during sleep)

- The upper and lower jaws not being ideally placed in the face

All of these conditions can lead to a person sleeping with their mouth open in a position that does not hold their airway open. Excess fat, the soft palate, tonsils, adenoids or the uvula can block the airway and cause those temporary stoppages in breathing.

The most common symptoms of sleep apnea are loud snoring and choking sounds. However, sleep apnea can occur with no symptoms – and even if there are symptoms, most people are asleep when they

occur and can be completely unaware of them. If you've been told you snore, and especially if you have unexplained trouble sleeping, feel tired during the daytime or suffer from headaches in the morning, you might want to take a sleep quiz to determine if you could possibly have sleep apnea. One such test is The Epworth Sleepiness Scale, in which you rate how likely you are, 0 being not at all likely to 3 being extremely likely, to fall asleep during the following situations:

- Sitting and reading
- Watching TV
- Sitting inactive in a public place (like a theater or a meeting)
- As a passenger in a car for an hour without a break
- Lying down to rest in the afternoon when circumstances permit
- Sitting and talking to someone
- Sitting quietly after a lunch without alcohol
- In a car, while stopped for a few minutes in traffic

If your score is high, that indicates that you should probably be evaluated for sleep apnea (unless there is some other factor in your life, like a new baby or sick child, preventing you from sleeping).

Treating Sleep Apnea

Once it has been determined that you do, in fact, suffer from sleep apnea, the next step is to treat it. Behavioral changes like losing weight or stopping smoking can help, but this is not guaranteed.

Many doctors prescribe to patients what is called a Continuous Positive Airway Pressure (or CPAP) machine – it requires them to wear a mask that blows air into their airway to keep it open while sleeping. As you might imagine, some people find this a bit uncomfortable.

This is where your dentist can be extremely helpful, as oral appliances that adjust your bite have proven to be a very effective alternative. These appliances work to keep your airway open by repositioning your jaw or tongue, and can be used to retrain you to breathe differently. They are usually fitted by your dentist specifically to suit your case and condition.

Beware of the generic "anti-snoring" devices you can buy online or at the store. As I've said before, you get what you pay for – since those appliances are not designed specifically for you, they can do nothing about your bite. Even worse, some can stop your snoring, but do nothing about your actual sleep apnea, leaving the most dangerous aspects of your condition untreated.

In some cases, the best option for treating sleep apnea may be surgery. Removal of the tonsils, adenoids and/or uvula can all be helpful if any of those organs interfere with your breathing. Oral surgery can also be an option, including repositioning your jaws properly in your face if improper positioning is a major underlying cause.

If you believe you may be suffering from sleep apnea, talk with your dentist about exploring your treatment options as part of your FMR.

8

Tooth Restoration

One important goal of Full Mouth Reconstruction is to make sure each tooth in your mouth is in the best shape possible – both from a functional and an aesthetic standpoint. So if your teeth have fractures, decay, wear, or dental work that needs replacing, addressing these issues needs to be a part of any FMR plan.

Unlike most conventional dentistry, however, FMR approaches all individual tooth problems with the final result in mind, by creating an overall game plan that addresses how all the different elements of the mouth are interacting and affecting each other.

Traditionally, most dentists in the past have just jumped in and done the restorations needed for individual teeth without taking the overall "big picture" into account. Most dentists, as a matter of fact, haven't even been trained to look beyond one or two troubled teeth – and they miss important opportunities to address the patient's overall situation and improve on it. It takes extensive continuing education on a dentist's part to enable him or her to go beyond the dental basics - and to be able to fully see and understand all the nuances involved in more complex cases.

An individual tooth that has work done to it may end up being just fine. And a patient can spend a lot of time and money over the years getting many individual tooth restorations done by the family dentist. The restorations may each be perfect; but they may not function well together or hold up over the long run, because they weren't done with an FMR perspective in mind. Just as periodic piecemeal reconstruction in a home usually leads to a crazy-quilt structure that could end up being unsafe, the final result of specific teeth fixes simply *cannot* be as good functionally and aesthetically, unless there is a customized and comprehensive FMR plan in place at the very beginning.

This chapter, then, will look at the various reasons individual teeth generally need restorations – and why you may want to consider a comprehensive FMR approach before you have those restorations done, especially if you have multiple teeth in need of them.

It may be in your best interest to at least *talk* with a qualified FMR dentist, who will give you a broader overview beyond your individual teeth problems.   It makes a lot of sense, since you'll need dental work either way, to look at the bigger overall picture – and get the procedures done in a manner that will enable your teeth to function the best and look the best for the longest amount of time possible.   After the FMR consultation, you can still obviously choose to do things in a less comprehensive manner - but you will at least have the knowledge you need to make the right, informed decision.

POSSIBLE REASONS TEETH MAY NEED RESTORATION

Tooth Decay

Do you remember having a cavity filled when you were a child or a teenager? If so, you've already had a brush with tooth decay. Decay happens when bacteria eat away at your tooth and form a cavity, or dental caries. It can also happen under previously-done restorations, where the restoration material meets the natural tooth - in fact, it's more likely that you will get a cavity under dental work than on a tooth that has never had a cavity before!

Cavities are often not painful and may not be visible without the use of x-rays to see between your teeth.

Fractures

No matter now strong your teeth are, they're still not indestructible – they can chip or break in certain circumstances. Even if a chip or crack doesn't bother you, repairing it may still be a good idea, and not just to make your smile look better. Chips and cracks can weaken teeth and make them more susceptible to bigger problems down the line. A fracture can serve as an entry point for cavity-causing bacteria. In addition, as fractures spread, they can affect the nerve (necessitating a root canal) or even split the tooth (possibly necessitating an extraction).

You can have a broken tooth and not know it. Pain is not always associated with fractures. Many times fractures don't become painful until just before they completely break or until they have reached the level of affecting the nerve. However, if a tooth hurts when you eat or expose it to heat or cold, that could be a sign of some kind of fracturing. The old silver-mercury amalgam fillings are exceptionally noteworthy in resulting in fractured teeth over time. This older filling material expands and contracts at a different rate than your tooth structure and is known to cause fractures over time.

Fractures can be very tricky. They are not always visible to the naked eye. They also are not often seen on x-rays. They can be very difficult to predict and very difficult to diagnose. It's also not often possible to determine the extent of the fractures prior to dental treatment. Therefore, if a fracture is evident, many times it is best to treat it.

Fractured teeth could be a red flag that you should consider consulting with an FMR dentist. It may be simple enough for any dentist to fix the individual tooth that has fractured - but what if the fracture was a result of your bite being off? Or what if, as a result of this fracture, your bite has now changed or shifted and it is now affecting other teeth in your mouth that you are unaware of? Multiple fractures could continue to result from a bite that's not right – that's why this is a case where an FMR dentist might make a big difference to your ongoing oral health.

Wear and Tear

Like most other parts of your body, your teeth suffer from wear and tear. Some people damage their teeth at a faster rate than others. Unfortunately, we only have one set of adult teeth. What used to last people only until they were of age for dentures, we are now trying to make last a lifetime – a lifetime that is much longer than in previous generations. Some people's teeth are more susceptible to wear than others, and at the same time, some people engage in behavior that is more likely to wear on their teeth.

When your teeth suffer from wear and tear, it can affect everything -- the way they look, the way they function, and even their lifespan. Depending on the severity, it can affect just your enamel (outer hard tooth structure), reach the dentin (softer underlying tooth structure), or even get to the nerves of your teeth. It's progressive and likely to get worse. Because your tooth isn't capable of regenerating, the damage is permanent – and once the wear reaches the level of the softer dentin tooth structure, the wear begins happening seven times faster than before!

Teeth wear also affects your outside appearance, making your nose and chin appear closer together – similar to cartoons you might see of older people. This condition can visibly make you seem to age faster.

Also, as your teeth wear down, it becomes much more difficult to do restorations on a single tooth. If a filling breaks, possibly as a result of the wear on it, a dentist can't rebuild the tooth properly because it still has to work with all the other teeth in your mouth – which are all still worn down. If you rebuild one tooth to its proper size and you don't fix the other worn down teeth, your jaw system won't work properly.

Generalized wear also becomes an issue if you want to restore your smile to that of your younger days. Perhaps you don't like how your front teeth look anymore and you think veneers would be a good option to make them bigger again, as well as fix those chips and fractures. The fact is, however, that your front teeth didn't wear down all by themselves. Usually, your back teeth are also affected. A dentist untrained in FMR may just do the veneers on your front teeth, which would result in them no longer functioning correctly with your back teeth. Your teeth may look great for awhile - but you could end up damaging or breaking them much quicker than either you or your dentist thought possible.

That's why it's important for your dentist to determine not only what can be done to fix your wear, but *why* your wear occurred in the first place. Often, there's more than one reason at work.

Causes of Teeth Wear and Tear

Attrition is the wearing down of your teeth's biting and chewing surfaces through contact with your other teeth. It affects the back teeth by flattening them, and the front teeth by shortening them. Most people experience this type of tooth wear to some degree – it can become a much bigger problem if your tooth structure has also been worn down by another cause, which will result in much more rapid wear. If your jaw system (your bite, jaw joints, and muscles) isn't functioning well together, this can also result in more rapid tooth wear.

A condition known as **bruxism**, involving tooth clenching and grinding, can cause severe attrition. If this type of grinding happens at night; a plastic or acrylic night guard can be worn over the teeth to help protect them. The problem is that the guard will only help while it's being worn; tooth wear will still occur when the guard is not in place.

Another form of attrition is known as an **abfraction lesion**. These are little cracks or splits in the enamel caused by the pressure of biting, chewing, clenching or grinding. These abfraction lesions occur along the gumline as little notches (or, sometimes, big notches) out of your tooth structure and may indicate that your bite is off.

Mechanical friction (such as tooth brushing) that wears away the surface of teeth is called **abrasion**. Tooth brushing can cause abrasion damage after the hard enamel of the tooth has been lost, possibly through another sort of tooth wear, and the dentin (underlying tooth structure) or root surface is exposed. Since dentin wears much faster (as noted earlier, up to seven times faster) than the enamel, this can lead to a serious situation.

When acids from food and drink or from your stomach eat away at your tooth surfaces, this is called **erosion**. All "fizzy" drinks (including regular and diet soda as well as carbonated mineral water), all sports drinks, and all fruit juices are acidic to varying degrees and can cause erosion. Pickles and citrus fruits are examples of acidic types of food that also can contribute to wear. Acid reflux and regular vomiting (bulimia) can cause severe erosion very quickly.

Why Is Dealing With Wear Important?

Tooth wear is a natural part of getting older. But it's still important to deal with it and treat it. The reason is simple: as tooth wear progresses, the entire mouth and facial dimensions change. It will even change the position of your jaw joints and may cause problems there. In order for your FMR to last a long time and function properly, tooth wear needs to be addressed, the lost tooth structure and vertical dimension of your bite may need to be replaced, and future wear needs to be prevented.

Extensive tooth wear should definitely be evaluated by a dentist trained in FMR. An FMR dentist will make the call as to whether it's a good idea to restore your worn teeth to their existing position, or if it's a better idea to change how your teeth and mouth are functioning as a whole. It may be in your best interest to reopen your bite and replace the worn tooth structure with porcelain to a better functional position. Neuromuscular dentistry may play a role in this type of treatment to make sure everything functions properly together.

Every situation is unique and needs to be evaluated from a comprehensive FMR standpoint to give you a smile that is healthy, functional and beautiful.

Replacing Old Dentistry

If you had any cavities filled when you were a child or a teenager, chances are it's time to have those fillings replaced. Many people don't know that fillings do not last forever. Chances are good that any filling you had done fifteen or twenty years ago is long past its expiration date.

When fillings stop functioning, it's difficult for the patient to tell. The decay eats away at the part of the tooth that's hidden by the filling, and often times, these conditions don't hurt… until it's a much bigger problem. Once you have symptoms like a toothache or a fracture, more serious conditions – requiring more serious treatment – will have developed.

If you have dentistry that was done prior to the 21st century, you most likely will have those black/silver amalgam fillings that were once the only filling material available. Amalgams tend to corrode over time, which lets bacteria sneak in through the broken seal. Plus, since amalgams contain mercury, they leak this toxic metal into your mouth. The American Dental Association says that amalgam fillings are safe, however, they also crack and break at the edges and are notorious for leading to cracks and breaks in teeth. In addition to treating the diseased condition of cavities and the structural issue of fracturing, there's a huge cosmetic benefit in replacing these old, amalgam fillings, as they're likely visible when you open your mouth to talk, laugh or eat.

When it is necessary to replace multiple old restorations it is a good idea to consult a dentist who can provide a full mouth reconstruction perspective. Old dentistry that is worn or broken can cause shifts in how your teeth fit together and how your jaw joints function. These changes may be minor, and, at first, may not affect your mouth function to any noticeable extent. However, if you are looking at replacing multiple old restorations, it may be in your best interest to have an FMR dentist evaluate your bite function and your teeth appearance to, again, ensure the best final result possible for your investment of time and money in having the work done. It's really the only way to achieve the best possible results for your dental health and the attractiveness of your smile.

Teeth Restoration Procedures

Now that we've discussed the different causes of teeth damage, let's look at the common restoration techniques used by dentists. Restorations are usually broken down into two categories: direct restorations and indirect restorations – so that's how we'll look at them in this section.

Direct Restorations

A direct restoration is one that is placed directly in your tooth and is shaped by the dentist during the process – this is usually called a filling or bonding.

Mild tooth decay can be treated by a filling; minor cosmetic improvements, such as fixing minor chips and correcting minor aesthetic issues, can be accomplished by what is sometimes called bonding.

In general, a filling is usually only appropriate if only a third or less of your tooth structure is being replaced. The benefits of doing fillings are that they are done in one sitting, they are a conservative treatment aimed at keeping as much of your healthy tooth structure intact as possible, and they are usually less expensive than other restorative options.

The negatives of fillings are that they usually don't last as long as other restorations, they are not as strong as other restorative options, they are usually not as visually appealing, and they can't be used to change or reposition your bite if that is a necessary part of your FMR. If the tooth in question is under a lot of pressure due to your bite, a filling might not even do the necessary job.

There are situations where a filling is the only option – or where it's just not possible. There are also, however, situations where a choice can be made. When that's the case, you should review the pros and cons of your options with your dentist to see what the best choice for you might be.

In general, fillings are made of either composites (tooth-colored plastic resin material) or amalgams (the old silver-colored, metal-mercury material). Because of its mercury content, amalgam's ongoing usage for fillings continues to generate much debate among dental professionals from both a safety standpoint as well as environmental concern. Some dentists don't even offer amalgam as an option any longer, and composite is a newer, safer and more attractive choice for most dental patients.

Feel free to discuss filling material options with your dentist, or you can even do a little online research on your own to find out more.

Indirect Restorations

An indirect restoration is created by taking exact measurements inside your mouth, creating a model of the tooth either in stone or through a computer program, and then custom-making it in a lab to fit on that model. Once it's made to fit on the model of your tooth, it can be placed on your actual tooth and cemented in place. Indirect restorations include crowns (caps), partial crowns (onlays and inlays), and veneers.

In general, anything that isn't considered a filling can be considered a type of crown. The term "crown" or "cap" was used in the past to indicate that the tooth being treated was completely covered by restorative material. The tooth was drilled down on all sides to make room for the crown material. Crowns were predominately made out of gold or porcelain-fused-to-metal (PFM).

More recently, it became possible to make crowns metal-free, using *only* porcelain. That technology meant that dentists no longer had to drill down a tooth on all sides – they only had to drill out the problem area of the tooth and replace only that area with the porcelain material. These partial crowns (which are also called inlays, onlays, or even veneers) leave much more of the healthy tooth structure intact and, therefore, are preferable when a full crown is not needed.

The benefits of indirect restorations are that they are made to last longer, be stronger, have a more attractive appearance, and can also be used to reposition or rebuild your bite, if that is a necessary part of your FMR treatment. The negatives are that they may take longer, they are usually more expensive and they may require the patient to wear a temporary restoration while the final one is being made.

Metal-free dentistry is today's state-of-the-art dentistry; porcelain materials can be used for many applications now. Everything from a small filling to a partial crown, a veneer, or a full crown can now be made from the same type of material. What once appeared as patchwork dentistry of different materials throughout your mouth can now be a unified and beautiful "look."

If your FMR treatment involves repositioning your bite and how your jaw joints function, you may have a transitional phase to your restorations. The decay will need to be addressed first, so that it doesn't progress - most likely with a temporary solution. You will also probably be placed in an "orthotic" or bite repositioning device to help reprogram your function and to determine the best functional position for you. Then, once everything has been worked out in the temporary phase to make sure the best function and esthetics are achieved, the final restorations will be built to match your new mouth position.

Which restoration is best for your situation will depend on many factors. Does the function of your bite need to be addressed first? Is there enough room for the various restoration materials in how your teeth fit together? How are the materials going to hold up if you use your teeth in their current functional state? What's the desired final result, both in terms of appearance and functionality?

While it's true that most people are looking for perfect white teeth that last a long time, you may first need to address some basic functional issues before that can happen. Again, a trained FMR dentist will be able to develop a viable long-range plan and explain your options to you in a way that will enable you to make informed, educated decisions.

Root Canals

This is really just a postscript to this chapter, as a root canal is technically not a "restoration" but a treatment that may need to be done in *conjunction* with a restoration.

A root canal is a procedure that is sometimes necessary if the nerve in your tooth is damaged. That damage may have many different causes – by a cavity that infects it, by a fracture that reaches it and causes pain, or by trauma (either by something immediately apparent such as an accident that injures your tooth or by less-obvious long-term trauma, such as your bite being off).

A root canal is also the punch line to many a comedy routine, whenever there's a reference to something extremely unpleasant and painful. In many cases, however, there is no pain involved when you need a root canal. It's very common to have a nerve in a tooth be damaged or even die without any accompanying pain.

Why? Because the affected nerve may suffer damage over an extended period, which allows the tooth to shield the pain. Sometimes an unhealthy nerve that isn't yet dead can be uncovered when dental

work is done on that tooth – then you can expect pain to result. In most cases, though, a patient will have no pain when the root canal procedure is done.

Prior to a tooth restoration procedure, it may be evident to the dentist that a root canal is necessary. It's not always possible, however, to know *before* the procedure is underway and the dentist actually uncovers a damaged or dead nerve. That means a root canal will have to be done or you run the risk of pain, swelling, infection, a failure of the restoration work or even loss of the tooth and surrounding bone.

So, even if you feel no pain or discomfort, if your dentist recommends a root canal, you should definitely get it done; as noted above, the consequences can be severe if you don't. The good news is that, despite the root canal's bad reputation, today's state-of-the-art technology means the procedure can be virtually pain-free. If you're still nervous about the procedure, you can always ask about sedation options.

9

Replacing Missing Teeth

If you have lost any teeth due to disease, an accident or other causes, replacing those teeth will be a key aspect of your Full Mouth Reconstruction. The goal of FMR is to help your teeth function and look their best, and to allow your jaw system for function optimally; this cannot happen if some of those teeth are missing from your mouth.

It is important to note that, as with other procedures performed as part of an FMR, the approach to replacing missing teeth will be different than it would likely be in the hands of a traditional dentist.

FMR dentists believe the best way to deal with missing teeth is to consider the whole picture first, then figure out the best way to treat your whole mouth so that everything functions well together. That means that if there is a problem with the way your teeth fit together, or how they function, or with the jaw joints and muscles involved, an FMR dentist will address those problems first rather than simply put replacement teeth back in the old – possibly non-ideal – positions. Making sure the mouth functions well as a whole means your replacement teeth will last longer and function the way they are supposed to.

A more traditional dentist probably would not look at your missing teeth quite that comprehensively. He (or she) might replace a tooth now, add a bridge later, and deal with any jaw pain in a completely separate series of visits at some point down the line. This could mean any underlying problems that may have contributed to the loss of your teeth in the first place will not be addressed, or it could mean that you will need to replace your replacement teeth on a more frequent basis – which can get expensive.

Why Do I Need to Replace Missing Teeth?

Tooth loss is very common among adults – between poor dental care, accidents and trauma, and the simple cumulative effects of old age, dentists often see people who have lost several teeth, or even all of them. However, as common as tooth loss is, it is important for people who lose teeth to replace them.

Appearance

The most obvious reason to replace missing teeth is the fact that, when you lose a tooth, it changes the way you look – and not for the better. A smile with visibly missing teeth looks unattractive, especially in an era where so many options are available for tooth replacement. And even worse, as tooth loss progresses, the entire face begins to change. Many people experience what is known as "facial collapse" -- when teeth are lost, the jawbone begins to recede, shortening the distance between a patient's nose and their chin. This can make a person look and feel older than they really are.

Confidence

The type of dramatic change in appearance caused by tooth loss can naturally lead to a loss of confidence. People with missing teeth often want to hide their smiles, because they fear what others might think of them. They may feel less comfortable in public, or meeting new people, or even around their friends and family. This could lead them to withdraw from social situations entirely, which can lead to depression and other health problems.

Health

This is where the impact of tooth loss is ultimately felt – in the rest of your body. You might not realize just how essential a good set of functional teeth is to good health; teeth are crucial.

Every tooth lost has a negative effect on your gums, your jaw and your other teeth. Your remaining teeth can move out of position, and the added strain on your jaw may lead to TMJ or other neuromuscular issues. People with missing teeth are also at increased risk for decay and gum disease as food and bacteria can become trapped in the open spaces.

A deteriorating jawbone and collapsing facial contours don't just affect your appearance; it affects your health as well. As your jaw function diminishes, so does your ability to bite and chew. This may prevent you from eating enough food, or the right kinds of food, to maintain good health. Foods like fresh fruits, vegetables, and whole grains, are essential to good health, especially as you get older.

The Three Types of Tooth Replacement

Living with missing teeth is never a good idea. If tooth replacement is delayed, the other teeth can shift and the jawbone can deteriorate enough so that replacement options become limited or not possible at all. It's important to have replacement options discussed as early on in the process as possible. There are several different ways to go about replacing them. The three basic categories are: removable dentures, fixed bridges, and dental implants. Each has its own pros and cons, as I'll explain here.

Removable Dentures

When you saw your grandmother's teeth resting in a glass by the bathroom sink, what you saw was likely a removable denture. Removable dentures have been around since at least George Washington's day, when he wore his famous "wooden teeth." Today, removable dentures are made of acrylic and metal and remain a common form of tooth replacement.

There are many different varieties of removable dentures, from inexpensive "dental flippers" (temporary false teeth held in place by a piece of plastic, usually used for only a short time until another more permanent replacement option is possible) to more substantial models made from metal and acrylic.

Those teeth in a glass you saw on Grandma's sink were likely what are known as *complete* dentures. These are used on patients who have lost all of their teeth; they cover the entire upper jaw, lower jaw, or both, and include artificial teeth and artificial gums. Complete dentures sit directly on top of the gums. Denture adhesive may be used to help keep them in place.

It is sometimes possible to snap removable dentures onto a few implants to help keep them in place. This type of implant supported removable denture is called an "overdenture" and is a major step up from a regular denture. An overdenture requires the placement of implants – implants are discussed later in the chapter.

A patient who has some healthy teeth left might wear a partial denture. These are held in place by anchoring onto the remaining natural teeth. Partial dentures usually include both artificial teeth and artificial gums, often with a metal framework used to connect the different denture sections and hold everything together and in place.

Removable dentures are based on old technology, and, with one notable exception I'll get to below, haven't really evolved much over the past few decades. The primary advantage of this type of tooth replacement is cost – removable dentures are less expensive than fixed bridges and far less expensive than implants. They can also be used in cases where a fixed bridge may not work because the span of missing teeth is too long (although implants are likely still a better option).

The disadvantages of removable dentures are numerous. They often don't look natural, and the wearer may be unhappy with his or her appearance. They can slip when the wearer eats or talks, making social interaction embarrassing for some people. They can be uncomfortable and feel unnatural in a person's mouth.

Additionally, while wearing a removable denture, the wearer's bite strength is a fraction of what it was with natural teeth – so eating normally will likely be a thing of the past. Some studies have indicated that half of denture wearers completely avoid many foods, and nearly 30% only eat foods that are soft or mashed.

Removable dentures can do nothing to stop the bone loss that naturally occurs at the site of a missing tooth. The shape of the jaw will constantly change, meaning the dentures will likely need to be replaced. Eventually the jaw may deteriorate to the point where it will no longer even support a denture at all.

This is only the beginning of the disadvantages of traditional dentures, which is why they are usually not recommended as a first choice for tooth replacement if other options are available.

Neuromuscular Dentures

Some people's only option for tooth replacement may be a denture. The good news is that there is now better technology that approaches denture design from a FMR perspective. This type of denture is known as neuromuscular dentures. Utilizing this new type of technology, a FMR dentist will take into account not only the function of the teeth and jaws, but also restore aspects of the face that may have been lost to facial collapse. The profile, the appearance of the lips, *everything* can now be addressed when dentures are designed using a neuromuscular approach.

While implants are now the preferred standard in tooth replacement for the best fit and function, and therefore the preferred tooth replacement method during FMR, sometimes they are not possible or sometimes a patient may need to choose a different option (for cost, timeline, medical necessities, etc). If a denture is going to be used, a neuromuscular dentist can use neuromuscular techniques to create removable dentures with a much better fit – which leads to much better function.

When neuromuscular principles are used in fitting dentures, the dentist focuses on the jaw muscles, making sure they are able to relax normally and function properly with the denture in place. The patient's bite is built at the exact position that will provide the best fit and function – enabling many neuromuscular denture wearers to eat a full range of foods again. Plus, since this position also provides the most facial support, patients who wear neuromuscular dentures also look younger and better than patients who wear ordinary dentures.

The detailed measurements necessary to properly design a neuromuscular denture take much more expertise than making a traditional denture – expect approximately four visits and expect to pay a higher price. However, since neuromuscular dentures fit, function, and look better, if you *must* wear a denture, isn't it worth getting the best design possible?

Fixed Bridges

A fixed bridge is essentially crowns connected together. Natural teeth on each side of the missing teeth are used as anchor crowns with fake crowns suspended between the anchor crowns, replacing the missing teeth area.

The advantage of a fixed bridge is that it represents a "step up" from the traditional removable denture. Fixed bridges are more stable; they look, feel and function more like natural teeth. More expensive types of bridges are almost indistinguishable from natural teeth (from an aesthetic standpoint), and dentists can even use a special technique, called the ovate pontic technique, to make it appear as if a false tooth is actually growing out of the patient's gum. Fixed bridges can also be used with dental implants, but we'll get to that later.

The primary disadvantage of a fixed bridge is that the teeth in the area of the bridge are all connected together. Another disadvantage is that if the anchor teeth are healthy teeth not in need of dental restorations, the patient essentially has to sacrifice healthy teeth to be able to support the bridge. In cases where crowns are needed anyway for the anchor teeth, a bridge is still a good option. There also may be situations where bridges are placed on top of implants. This is often the case if too many teeth in a row are missing; implants are used to replace a couple of the teeth and a bridge is made that is placed on the implants. A fixed bridge supported by natural teeth isn't feasible in cases where too many teeth in a row are missing or if a back tooth isn't present to provide an anchor.

Dental Implants

Dental implants are the current state-of-the art technology in tooth replacement; they are the most natural functional and aesthetic choice to replace missing teeth. Dental implants look and function closest to natural teeth – except for the added plus that they don't get cavities. In fact, depending on what their teeth looked like beforehand, many patients consider their implants better than their natural teeth.

However, an implant alone does not look like a tooth at all. A dental implant is really just an artificial tooth "root" that is surgically implanted in a patient's jaw. It looks like a metal screw. The visible "tooth" part can be a crown, bridge, or denture that is attached to this artificial root.

The advantages of dental implants over other forms of tooth replacement are numerous. In addition to looking and functioning like natural teeth, the metal used in dental implants, titanium, is used because it actually fuses with your jawbone and becomes a part of your body. This fusion, called "osseointegration," stops bone loss in the jaw, preventing facial collapse, and keeping implant patients healthier and stronger than their denture-wearing counterparts.

Implants have no affect on any healthy teeth you might have in your mouth. No teeth need to be "sacrificed" to support a dental implant, so you can retain whatever teeth you have for as long as they remain healthy. Additionally, implants can be configured to span as much space as you need, from a single tooth to an entire mouthful.

Still, no tooth replacement is "perfect," and implants do have some disadvantages. The primary downside is cost – implants are much more expensive than dentures, and dental insurance rarely covers them. Getting implants also takes time – a total of several months and multiple dental visits, although you never need to be "without teeth" during the implant process; it takes time for the implants to integrate into your jaw bone before the final, permanent crowns can be placed.

Getting dental implants is a surgical procedure. Many FMR dentists will partner with a surgeon to place the implant part of the restoration; usually either an oral surgeon or periodontists. It is important that your FMR dentist communicates in detail with the surgeon about the final goals of what the implant will be used to support, to ensure that they are placed in the most ideal position for the planned final outcome. Not everyone can have dental implants – enough bone of good quality must be present to support the implant and any chronic conditions like diabetes or heart disease need to be under control.

A person needs to have enough healthy bone to support an implant. In cases where bone has been lost, a surgical procedure called "bone grafting" may be used to place healthy bone or artificial bone where it is needed to enable implant placement.

If you are missing teeth, know that replacing them will be a major part of your Full Mouth Reconstruction. Your FMR dentist will work with you to find the solution that works best for you.

10
Cosmetic Dentistry

Unlike a so-called "Smile Makeover" that is really only focused on how your smile looks, a Full Mouth Reconstruction is concerned with how your smile *functions* – so FMR includes treatments to make sure your teeth, gums, mouth and chewing system are healthy and work properly.

That doesn't mean how your teeth look is not a prime concern of your FMR dentist. The ultimate goal of an FMR is to give you a smile that will look just as dazzling as, if not better than, a "smile makeover" smile, with the added bonus of health, durability and proper function. If a smile that is whiter, more even, or shapelier is your goal, cosmetic dentistry will be an important aspect of your Full Mouth Reconstruction plan.

Even if you are certain that you only want a "smile makeover," seeing an FMR trained dentist still could be your best option. First of all, it can't hurt to hear the perspective of someone who has this kind of advanced training. Second, if there is an underlying functional problem you don't know about -- a leaky pipe that is making your plumbing fail faster, or a hinge out of line that is making your door wear unevenly -- you should be aware of it before you decide on a treatment plan.

When you solicit the opinion of an FMR trained dentist, you're giving yourself extra insight into your dental health that will help you make the best decision for yourself. Even if you ultimately decide that you only want the smile makeover, a dentist who understands the full mouth perspective may give you more functional (and possibly more attractive) restorations than the "cosmetic" dentist down the street. Don't be afraid to use an FMR dentist who has recommended treatments you have decided against. As long as he or she knows that you have been educated and understand what your choice means, an FMR dentist is likely to provide the treatment you want.

Those treatments may include the following:

Braces

Most people think of orthodontic treatments like braces as a way to improve their appearance – which is why I've placed braces here, in the Cosmetic Dentistry chapter. But the fact is, if your teeth are crooked, or are not lined up evenly and in the right places, chances are they're not functioning at their best either.

An uneven bite can lead to tooth wear, TMJ problems and even sleep apnea. So as you might imagine, I prefer to look at braces from an FMR standpoint. That means first asking the question, "How can orthodontic treatment improve the appearance, function, durability and health of this patient's smile?"

What are Braces?

You may have had braces as a child, or at least knew someone who had them. In case you are not that familiar with them, dental braces are used to bring a patient's teeth into alignment. They are used to treat conditions known as overbites, underbites, cross bites, open bites (when the teeth don't close all the way), deep bites (when the teeth sink in too far) as well as crooked teeth.

Braces move teeth by applying consistent pressure. This is accomplished using different types of devices, including:

> Traditional Braces – These are metal brackets that are attached to the front of the teeth; a support wire is tightened gradually to move the teeth into position.

> Clear Braces – These use ceramic or plastic for the brackets, which makes them less visible.

> Gold-Plated Braces – These are used for patients who are allergic to the metal nickel, which is used in traditional braces. Some people also choose gold braces because they like the color.

> Lingual Braces – also called Incognito Braces; these have brackets attached to the back of the teeth so they cannot be seen.

"Invisible Braces" or Aligners – These differ from ordinary braces because they can be removed for eating and special occasions. Each see-through aligner is worn for a period of a few weeks, then replaced with a new one, to guide the teeth into position. They are not usually ideal for complicated cases. With this type of orthodontic treatment, the results are dependent upon patient compliance in wearing the aligners ALL of the time except when eating.

The "Smart Bracket" – This is a new technology that may or may not be available in your area, using a computer microchip in the bracket to optimize the amount of tension applied to teeth and speed up orthodontic treatment.

A-braces – another new form of braces; these are completely controlled by the user, adjustable and removable, so the patient can control the amount of pressure felt. Again, the results achieved are dependent upon patient compliance.

Because of the pressure they apply to your teeth and mouth, and because they take up additional space inside your mouth, wearing braces can be uncomfortable. In some cases, they can even be painful, although the pain can usually be treated with an over-the-counter pain reliever and usually subsides in a day or so.

What to Expect

If your FMR dentist feels that orthodontic treatment would be beneficial, he or she may personally perform it or refer you to an orthodontist. The X-rays, molds, and impressions taken of your teeth for your FMR will be analyzed, and the right treatment option selected. Expect treatment to last from six months up to two years or more, depending on your individual case.

Sometimes, there is not enough space in the mouth for all of the teeth to fit properly, which can lead to crowding. Some dentists or orthodontists might suggest having teeth removed to prepare your mouth for braces, but this approach is generally not ideal. If possible, a better option is to expand the arches, which can create space for proper function and aesthetics. In more extreme cases, surgery may be recommended.

FMR dentists understand that removing teeth (other than the wisdom teeth or 3rd molars) can result in an imbalance in your jaw system and facial structure, which can have a negative effect on your appearance. This being said, there are some cases where removal of a tooth or teeth may still be the best option for a particular case, but it tends to be the last resort.

Most types of braces (excluding "invisible braces" aligners) are typically attached to the teeth with adhesive – although in some cases, bands are also used to hold the bracket on the tooth. A strong wire called an "arch wire" is threaded through the brackets, and that wire is shaped and tightened to pressure the teeth into position. Depending on your particular situation, you may be instructed to wear elastic bands or additional, removable appliances.

Once your braces are in place, expect to visit your orthodontist regularly – at least every month or two - to have your wires adjusted and tightened. As this is when pressure feels the strongest, you may experience some discomfort after these appointments.

The Right Orthodontics for FMR

The kind of orthodontic treatment your FMR dentist will recommend is likely to be different from the braces you wore as a child. As I mentioned earlier in this book, FMR is a new approach to dentistry, looking at the way the entire chewing system functions as a whole instead of treating problems piece by piece. So it only makes sense that if braces or orthodontia are going to be a part of your FMR, the process will be about more than giving you "straight teeth." The right orthodontist will treat the causes, rather than just the symptoms, of your bite problems and do it in a way that focuses on your whole body, not just your teeth.

Why is this approach so much better? When a typical orthodontist sets out to give you the straight teeth you want, he or she usually accomplishes this by pulling your teeth back in your head. This can change your facial structure, and not always for the better. An orthodontist who understands FMR principals will utilize different methods focused on maintaining the balance in your face as well as proper function of your jaw joints and your airway.

Orthodontic problems are often caused by what is called "poor oral posture," when the tongue, teeth and lips are not in the right position when the face is resting. An orthodontist who understands FMR will work to correct those problems, so your orthodontic treatment will look beautiful and last a long time.

Of course, if your only problem is crooked teeth, this type of orthodontist can correct that too.

Tooth Whitening

Whitening teeth is the most common cosmetic procedures available. It can produce dramatic results in many patients. While it serves no real functional or health purpose, it can usually be included in the full mouth reconstruction process to help teeth that are discolored become more attractive.

For some people, it is not possible to achieve the desired amount of color change with whitening. In cases where tooth discoloration is due to wear, aging, or chemical discoloration, whitening may not be the right treatment to produce the desired result – there are other ways to improve these teeth, which we'll get to later. In addition, whitening is not recommended for pregnant women.

You may have heard tooth whitening referred to as "bleaching." Keep in mind that this term can only be used when the treatment can make teeth even lighter than their natural color, and when the products used contain a bleach like hydrogen peroxide or carbamide peroxide.

"Whitening" can refer to any treatment that removes staining and makes teeth look whiter – which is why some toothpastes can be called "whitening" toothpaste. Still, many dentists use the term whitening instead of bleaching because they think it sounds better. Be sure to ask your dentist specifically what treatment they are recommending for you and what range of shade improvement you can realistically expect to achieve. Results can vary widely from person to person.

How Whitening Works

Most teeth start out fairly white, but, over time, the enamel on the surface wears down, letting more of the color of the yellow dentin below show through. While this is happening, stains from food and beverages enter tiny cracks in the enamel and stick to the teeth, discoloring them.

"Extrinsic stains" are the kind of stains caused by beverages and foods like red wine, coffee and berries, as well as tobacco use and regular wear and tear. These are stains on the outside surface of the teeth and are the easiest types of stains to treat.

"Intrinsic stains" actually form on the interior of teeth, and are caused by exposure to minerals (like drugs or excessive fluoride), aging and trauma. These stains are much harder to lighten, but can sometimes be improved through professional, in-office treatment combined with professional in-home teeth whitening that takes place over several months.

Keep in mind that whitening only lightens the color of your teeth. If your teeth are worn, crowded, or fractured and discolored, if you whiten them… they will still be worn, crowded, or fractured, albeit a few shades lighter.

Another option with difficult to whiten teeth or where whitening isn't the only cosmetic concern is to consider dental veneers, which we will get to later in this chapter.

Why Do Teeth Stain?

There are several factors that contribute to tooth staining, including:

Age – the older you get, the more likely your teeth are to stain as they experience more wear and tear and stains accumulate over time.

Natural tooth color – the natural range of tooth undertone shades range from yellow to brown to grey. These colors get more intense over time. Teeth with a yellow undertone respond the best to whitening. Brown undertones tend to get moderate improvement, and grey undertones are the most difficult to treat.

Translucency and thickness -- Thicker, more opaque (less see-through) teeth stand up to age better and respond better to bleaching. Teeth that are more translucent don't respond as well to tooth whitening.

Eating and drinking -- If you love your red wine, coffee or tea, chances are your teeth will show it. Acidic foods like lemon juice and vinegar also erode tooth enamel, allowing the yellowish dentin to show through.

Smoking – nicotine can stain teeth intensely.

Drugs / chemicals – the child of a mother who took Tetracycline while pregnant may have teeth that are stained grey or brown. These stains can be very difficult to treat and may be better treated with veneers. Consumption of too much fluoride can also stain teeth with mottled, white spots. Mottled white spots can often be treated well with a remineralization process prior to whitening.

Teeth Grinding – the tiny cracks caused by grinding attract and hold stains.

Accidents and trauma – these can also produce cracks, which attract stains and debris.

Nerve damage – if a tooth has nerve damage either from trauma or decay, often the color of the tooth will darken internally. This type of darkening can often be helped with a root canal and internal bleaching of the crown of the tooth.

Types of Teeth Whitening

Three major teeth whitening options are available today. They are:

- ## In-Office Whitening

In-office whitening treatments can do a lot in a short amount of time – often only a single office visit. During the procedure, a dentist or trained assistant will apply a highly concentrated peroxide gel to the surface of your teeth. The gel typically remains on the teeth for a few periods of 15-20 minutes adding up to an hour or less. Patients with stubborn stains might need to return for a follow-up visit, or may be advised to continue with a home-use whitening system. Often, the home trays will be recommended to maintain your new whiter smile, providing you with the ability to do "touch up" treatments at home.

- ## Professional Take-Home Whitening

These kits, which are prescribed by dentists and sent home with their patients, utilize a more "slow and steady" approach to tooth whitening. The patient is given custom fitted whitening trays, along with a gel that uses a lower concentration of peroxide than the "in office" version. Dependent on the strength of the bleach given to you to use at home the time may vary from 30 minutes to overnight.

- ## Over-the-Counter Whitening

If you're undergoing FMR, it's highly doubtful that your dentist will suggest one of these inexpensive, less-effective kits to finish the job. Store-bought whiteners have a lower concentration of peroxide, and some varieties only whiten a handful of front teeth.

Teeth Whitening Risks

Teeth whitening is generally safe, but it is not for everyone. Even patients who are well-suited to whitening can experience increased sensitivity to hot and cold, and possibly touch as well. Some patients even experience shooting pains in their front teeth. Since in-office whitening uses the strongest concentration of bleach, sensitivity is most likely to occur after this procedure.

Sensitivity is more likely to affect patients who have recessed gums, large cracks in the teeth or faulty restorations – something no FMR patient should have, as these problems will likely have already been addressed. If you should still feel some sensitivity, it typically only lasts for a few days and over the counter pain medications can help alleviate it. To reduce the occurrence of sensitivity, you can ask your dentist about using a desensitizing gel for a period of time prior to the whitening appointment.

Many patients also experience gum irritation, which tends to wear off after either at-home treatments have stopped or a few days after an in-office treatment.

Keeping Whitened Teeth White

Maintaining your new white smile may require a little extra maintenance. In addition to regular dental care, your dentist may recommend avoiding dark-colored foods and drinks, especially for the first week or maybe longer. There may also be some sort of "maintenance" or follow-up whitening plan that can range from at-home treatments to additional treatments in your dentist's office.

Be sure to ask what maintenance will be involved when you discuss whitening options with your dentist. Together, you'll come up with a plan that works for you.

Replacing Old Dentistry

I discussed replacing old dentistry in a previous chapter. This can be considered "cosmetic" dentistry as replacing the older materials that are not especially attractive with the new, natural looking, tooth colored restorations can have a huge cosmetic benefit. Even the older "white" restorations don't have a very natural look to them; the newer materials are far superior aesthetically and functionally.

Dental Veneers and Bonding

Dental veneers and bonding are cosmetic procedures used to improve the appearance of individual teeth, a group of teeth, or an entire smile. What type of treatment will work best for you depends on the extent of the issues you want to have addressed.

Bonding

When it comes to changing the look of individual teeth, bonding is usually the most inexpensive option. However, it is only really appropriate for minor issues like small cracks or chips in teeth. During the bonding process, a tooth-colored, composite resin material is "bonded" to the tooth to fill up the crack or chip and protect the remaining tooth structure. This is a simple, non-invasive procedure that can be done in a single office visit. The downside is that bonding doesn't generally look as good or last as long as veneers, and it is more prone to staining. In cases where the function of your bite or how your teeth fit together is the cause of gradual chipping or wear, don't expect bonding to hold up very well.

Veneers

If one goal of your FMR is to dramatically change the look of your teeth, dental veneers may be your best option. They are most often made of porcelain and are long lasting. Veneers can be used to correct a wide variety of aesthetic issues, to mention a few:

- chipping and cracking

- dullness and discoloration

- wear

- size issues

- problems with tooth spacing, crowding, or rotations.

In some cases, crowns made of the same porcelain material as veneers are used for functional purposes like protecting a damaged tooth or replacing a crown.

Veneers are, essentially, a custom made, tooth-colored "shell" that fits over the surface of a tooth, improving its color, shape and even position in your mouth.

There are different types of veneers available:

- **Traditional Veneers**

These are usually made of porcelain or composite resin – the porcelain variety is longer lasting, better looking, and more expensive. Both types can be fabricated in a lab outside your dentist's office; however, composite veneers can also be created in your mouth by your dentist. Veneers that are fabricated in a dental lab are bonded to your teeth by your dentist with resin cement.

Traditional veneers typically require the removal of some of your tooth structure through shaving or grinding it away. Since this process will alter those teeth permanently, traditional veneers may not be recommended for teeth that are already attractive and functional. In cases like these, other options, such as orthodontia to correct spacing issues, may be recommended instead.

Preparing teeth for traditional veneers usually requires the use of local anesthetic. If you are a fearful or sensitive patient, you should ask your dentist about sedation options.

- **"No-Prep" Veneers**

"No-prep" veneers are less invasive and require no or minimal tooth removal and reshaping. There are several widely marketed brands including Lumineers, Durathin and Vivaneers. These brands are simply companies trying to get you to request their company by name. Just as with traditional veneers, the no or minimal prep veneers are most often fabricated from porcelain. If you are interested in one of these marketed brand names, you should discuss the pros and cons with your dentist -- often it is possible to use these products, but they may not be the best option for your case.

Minimal to no prep veneers are bonded to the front of the teeth like traditional veneers. In the cases where minor reduction is necessary, it usually only involves the enamel. As an added bonus, in some cases no anesthetic is required (although it is usually available to nervous or sensitive patients).

While minimal and "No-Prep" Veneers are not suitable in many cases, there are cases where they can be successfully used, including:

- Chipped, cracked or worn teeth

- Slight discoloration or staining

- Spaces between teeth or slight crowding

- Small, misshapen or slightly misaligned teeth

Since all teeth are different, there are many cases when some teeth require more work and others need only minimal work to achieve the desired result. In cases like these – and there are a lot of them - - a combination of traditional and minimal to no-prep veneers can be used. A good FMR dentist will let you know if a plan like this is right for you, or if other options – such as braces or aligners to straighten teeth or whitening to lighten them – might be a better option.

Which Veneers Are Right For You?

What type of veneers will work best for you will depend on three factors:

> Your existing teeth (how they are positioned, shaped, colored, what dentistry is already present, any decay, etc.)

> What you want or need to change about your teeth

> What the final goal is (how your final teeth are to be shaped, positioned, colored, etc).

Once you and your dentist have considered these factors, your dentist will come up with the right veneer plan to get your smile from point A to B.

For example, if your teeth are small, have spaces between them, are not too dark and are generally aligned well in your mouth, it will be easy for your dentist to simply cover your teeth with minimal to no prep porcelain veneers to achieve a beautiful, even, white smile.

On the other hand, if your teeth are dark, rotated, crowded together or sticking out in places, your dentist will probably need to take away some of your existing tooth structure to make room for the veneers. Otherwise, you will wind up with teeth that are too big and bulky, and you may not wind up with a natural-looking color variation. If you have old dentistry, it will usually need to be removed so as to not cover up any hidden problems.

As with most things in dentistry, every case is unique and every tooth position and shape is unique. It's possible to have 5 teeth that need no preparation for veneers and 2 that require work. It's also possible to only need a single veneer, to cover one tooth that has been discolored or otherwise damaged.

It's best to approach your dentist with an open mind. Rather than ask for a particular type of veneer, talk with your FMR dentist to about the options that will work best in your particular case. That way, your end result will look better and last longer.

What to Expect

The process of veneers varies, depending on what type you are getting – these are the general steps you can expect with traditional veneers.

> Smile design – discuss what your goals are with your dentist and come up with a plan.
>
> Design the teeth on models
>
> Prepare the teeth for veneers – any tooth reduction would be performed here, along with anything else that might be needed to get the teeth ready for the desired final results.
>
> Impressions of prepared teeth – if veneers will be fabricated in a lab, impressions will serve as a model that the veneers will be fit to.
>
> Temporary "trial" smile – If your veneers are being fabricated in a lab, you'll receive temporary restorations in the shape of your desired new smile to wear until your final

veneers are fabricated. The color may not be perfect because temporary materials come in a limited range of shade options. However, the temporary veneers will likely be close to the desired shade, very close to the planned shape, and will likely be a huge improvement from what you walked in with.

The temporaries will provide you with your first opportunity to see your new smile before your permanent veneers are placed. I call this a "trial smile" – because during the waiting period (which I'll get to next) you'll have time to get used to your new look and see if you want any changes before they are finished in the lab.

Waiting period – It usually takes up to three weeks for the lab to custom build your new smile. This might seem like a long time, but it's worth the investment to make sure you have time to evaluate your temporaries and the lab has the time to create the beautiful results.

Placement appointment – the big day when you finally "get" your beautiful new smile!

Adjustment/final check appointment – You'll return at least one more time to make sure that your bite is right, that no residual bonding/cement material is left over, and that you are able to properly care for your new smile. At my office, this is also when we take your final photos, because your tissues will have had time to adjust to your beautiful new teeth.

If your FMR also includes having the position of your bite changed, you may need several appointments to adjust how your new veneers are functioning with your new bite.

Caring for your Veneers

Dental veneers are designed to last. However, the porcelain shells can break or sustain damage just like natural teeth. Your dentist will tell you if you need to be careful about what you eat, and if you grind your teeth, he or she may recommend you wear a protective appliance while your sleep.

After all, you'll want your beautiful new smile to stay beautiful for years to come.

11
Sedation

According to the American Dental Association, up to 80% of adults – that's four out of every five people – have experienced at least some fear of dental treatment. Up to half of those people (40% of American adults, if you're keeping track) admit that they avoid the dentist because of these fears.

Maybe you're one of those patients. Maybe one of the reasons you're considering a Full Mouth Reconstruction now is because you were too frightened of the dentist to get the dental care you needed when you first needed it.

On the other hand, maybe you've always done what you needed to do to keep your smile healthy, but struggled with anxiety or pain, or both, during your appointments.

Or maybe you've never had a problem with the dentist, but are feeling nervous about the more intensive treatments that might be a part of your FMR.

Regardless of the reasons behind it, if you feel at all anxious about any or all of the treatments that will be part of your FMR, you don't need to be. As dental treatments have improved and evolved over the past few decades, so have many dentists' attitudes towards and methods of managing their patients' fears and preventing their discomfort.

The result is that pain and stress-free dental treatment is now a possibility for even the most fearful or sensitive patient, thanks to Sedation Dentistry.

What is Sedation Dentistry?

Sedation Dentistry is essentially a catch-all term for dentistry performed with a special emphasis on patient comfort. For example, instead of simply giving you a local anesthetic injection (such as a Novocain derivative) before treatment where pain might be involved, a dentist who practices sedation dentistry techniques will discuss your fears with you ahead of time and ensure your comfort with additional medications an integral part of your treatment plan.

When used properly, sedation ensures that even the most involved treatments that might otherwise lead to some level of discomfort don't cause any pain, and that anxiety you might experience before treatment begins is alleviated. Additionally, some types of sedation make it seem like your dental treatment only lasts a few minutes, which makes it ideal for complex procedures and may help your dentist get more done in less time.

Why do People Need Sedation?

Everyone is different: therefore, there is no one reason why a patient might want to use sedation as part of their dental treatment. With that in mind, here is list of those fear factors that bother people the most:

- **Fear Of Pain During Treatment**
  The most common -- but by no means the only – reason for dental fears.
- **Fear Of The Dentist**
  For some people, the mere sight of a medical or dental professional in a white coat can be terrifying.
- **Embarrassment About The Condition of Your Teeth**
  This can include the fear of being scolded or reprimanded by the dentist as well as embarrassment at showing your teeth to anyone.
- **Fear of Lack of Control in the Dentist's Chair**
  Some people actually fear they'll receive the wrong treatment - or treatment they don't want!
- **Fear of Needles**

People who are especially uncomfortable getting injections often experience dental anxiety.

- **Fear of the Atmosphere of the Dental Office – Sounds, Sights and Smells**
  These are sometimes associated with negative past experiences a patient may not even remember.
- **Fear of Gagging or Choking**
  Some people are frightened by the idea of not being able to swallow.
- **Fear of Crying or Having a Panic Attack**
  People who are especially fearful may worry about the reaction that fear will provoke.
- **Fear of The Drill (or other Dental Equipment)**
  The sight and sound can be enough to terrify some people.
- **Issues With Numbness**
  Some people worry they won't be able to breathe when they are numbed, while others worry they won't be able to get numb in the first place.
- **Fear of Being Awake**
  Some people fear being fully aware during treatment and being able to (or being forced to) fully experience everything that is going on.

Any or all of these fears can make the basic steps necessary to taking proper care of your teeth, let alone more involved procedures like those in an FMR, especially difficult. That's why Sedation Dentistry was developed – to provide fearful patients with a real solution to any level of dental-phobia, so that visiting the dentist is painless, stress free, and in some cases, even pleasant.

How Today's Sedation is Different

Sedation has always been used in dentistry to some degree. But today, what we call "Sedation Dentistry" is a bit different from what you might have experienced when you had cavities filled as a child. That difference primarily centers on the situations where sedation treatment is offered, the types of sedation, and how the sedation is administered.

In the "old days," dentists restricted their use of sedating medications to reducing or eliminating pain. Sometimes, a little pain was able to sneak by, which, in a fearful patient, could constitute a Bad Experience that might lead to more fear.

There are, of course, still plenty of dentists who practice that way today, paying no more attention to their patients' comfort than they did twenty years ago.

If you are a nervous or uncomfortable patient, this is definitely not the kind of dentist you want performing your FMR. However, a dentist who still practices in the Dark Ages is not likely to be on your list of potential FMR providers.

So the important thing to remember here is to communicate with your dentist before your treatment begins. Share your fears and anxieties – believe me, as a dentist, we've heard it all before and we're not going to judge you. As long as you make dealing with your comfort a priority, your dentist will too. He or she just needs to know the extent of your sensitivity or fear, so that the right plan can be created to address your needs.

Types of Sedation

Today's sedation options have come a long way since the first patient inhaled nitrous oxide (which you might know as laughing gas) in the 1840s.

Dentists today can approach sedation with the ultimate goal of completely relaxing their patients so that they can have the dental treatment that they need without feeling nervous or uneasy. This is accomplished with the use of other types of drugs including tranquilizers, anti-anxiety drugs, and depressants, administered in a variety of ways.

The goal ranges from mild anxiety relief to a deep sense of relaxation that can make patients less aware of what's going on around them – including pain. In extreme cases or for major treatments, some dentists even offer the same kind of general anesthesia you might experience during medical surgery.

Inhaled Sedation

The oldest type of sedation, inhaled sedation is still commonly used
– especially at more "old fashioned" dental offices. During
sedation, the patient wears a device that enables and gently forces
nitrous oxide into their airway. Nitrous oxide can make patients
feel mild euphoria, which is why it is sometimes called "laughing
gas." It is not, however, particularly strong, so it may not be
enough on its own to block pain or ease anxiety in fearful patients.

Oral Sedation

During oral sedation, the patient is given a strong relaxing or
tranquilizing drug in pill form and advised to take it a specific
amount of time before their appointment. The medication in this
type of sedation is typically stronger than that in nitrous oxide, so
the patient will likely relax more and feel and remember less. As an
added benefit, no needles are required for oral sedation, which can
be an extra help to anxious patients.

Intravenous (IV) Sedation

A strong relaxing or tranquilizing agent is given to the patient
through the blood, via an IV tube inserted into their hand or arm –
however, the act of being injected may make some patients nervous.

The stronger the sedative used, the stronger the possibility that you
will feel sleepy during treatment, and that you might not remember
it. However, as you will be *technically* awake, the term "sleep
dentistry" is not accurate and should not be used.

Is Sedation Right for Your FMR?

If you're one of those people who is nervous or uncomfortable
going to the dentist, sedation will help make the procedures during
your FMR much more pleasant. However, even if dental visits
don't generally bother you, you might want to consider sedation
during your FMR, as it makes time to seem to pass more quickly

and will enable your dentist to get more done before you are too tired to continue. .

Specifically, Sedation Dentistry can be especially helpful to people who:

- Have been affected by a past, negative dental experience

- Are afraid of experiencing pain at the dentist

- Have a fear of needles or drills

- Have sensitive teeth or gag easily

- Have avoided the dentist to the point where they have serious dental problems that need attention

- Want to take care of multiple dental treatments quickly and get as much done in a single visit as possible

- Simply want a more relaxing, comfortable experience in the dentist's chair

For these types of people and many others, sedation can make Full Mouth Reconstruction much easier and significantly reduce stress.

There is one note of caution. If you are going to undergo sedation during any of your procedures, make sure you've made arrangements for a responsible person to drive you home, and have allowed yourself enough down time for the sedatives to wear off. Sedation is safe and effective, but real drugs are involved, so some caution must be taken when it is used.

12

Maintaining Your New Smile

Once your Full Mouth Reconstruction is complete, you'll have a beautiful, healthy, functional smile that you'll want to show off to the world. However, while a good FMR dentist will definitely design your smile to be as durable as possible, how long your new smile stays healthy, functional and beautiful depends on you. FMR is only the first step in the process of restoring your dental health. The next step is maintaining it.

As you might expect, that means starting with proper dental care. It means brushing at least twice a day, and possibly more, depending on the type of work you've had done on your teeth. It means flossing, possibly using a waterpik, or implementing any of the other homecare regimens your dental team recommends. It means seeing your dentist as often as recommended for regular checkups, cleanings and any necessary in-office maintenance.

More importantly, it could mean changing the way you've taken care of your teeth in the past, especially if your Full Mouth Reconstruction included any restorations or veneers. The materials used for dental restorations are designed to look just like natural teeth, but they often wear differently. Therefore, it is important to understand that you may need to change your habits in order to protect your investment in your dazzling new smile.

Most restorations used in FMR will be made of a bonded composite resin, porcelain, or some combination of both materials. They are designed to be durable, however, like natural teeth; they can also stain, chip, or even break. To prevent that from happening, you will want to keep this list of Do's and Don'ts in mind.

FMR Maintenance Do's and Don'ts

DO – Attack Plaque

Controlling the level of plaque on your teeth (both natural and artificial) is essential to keeping your smile healthy – which means regular brushing and flossing is a must. Plaque can wreak havoc on the materials used in dental bonding, making them porous and more susceptible to internal staining. It can also cause gums to recede, exposing parts of teeth that might be a different color than, say, a gleaming white veneer. Recession can also cause increased tooth sensitivity to temperature changes and sweets. Increased levels of plaque will also lead to gum disease and associated problems (see the earlier chapter on gum disease).

As a result, FMR patients who let plaque get out of control typically need to replace bonded fillings and veneers more frequently than patients who keep plaque in check.

When you brush, pay special attention to the area between where your dental work stops and any natural tooth or gum begins. This is probably the most susceptible to bacteria – the tiny spaces are ideal places for them to "set up camp."

DON'T – Chew Hard Things

Maybe when you had your natural teeth, you liked to chew on ice, or used your teeth to tear those hard-to-open plastic packages or bite down on pins or staples. Maybe you had a nervous habit like biting your nails or chewing pencils. If you did, it's now time to stop.

Habits like those can be bad for your teeth, but they're even worse for dental restorations. They can cause teeth to chip, crack or break, and if you've just invested a lot of time and money in a beautiful new smile, one bad bite can ruin your investment amazingly quickly.

Eating normal foods like bagels, apples and corn on the cob should be fine, although some hard foods can harm bonded restorations. So use common sense when using your new teeth.

DO – Whatever It Takes To Stop Grinding Your Teeth

Bruxism and tooth grinding weren't good for your teeth before your FMR – and as you might imagine, they won't do your new, improved smile any good either. The constant pressure placed on restorations by grinding can cause them to chip, crack or break.

Of course, chances are good that if you have this problem, your FMR dentist has already spotted and addressed it as part of your treatment. If that is the case, be sure to faithfully wear whatever type of night guard your dentist has suggested to protect your new smile.

DON'T – Drink Alcohol to Excess

This "don't" surprises a lot of people – after all, what does enjoying a few alcoholic beverages have to do with your teeth? The key here is how many drinks constitute a few. While dinking a moderate amount of alcohol doesn't appear to affect dental restorations, having several drinks a day softens bonding material over time, affecting both direct bonding and the bonding used to hold veneers in place. If the drinking is consistently heavy, it can ruin your restorations in less than two years.

Even if you don't drink, you could still be exposing your new, improved teeth to too much alcohol if you use a mouthwash that contains it. So if you do use mouthwash, check the ingredients, and if alcohol is on it, choose another brand.

DO – See Your Dentist Regularly – And Probably More Often

Veneers and other restorations can sometimes benefit from extra maintenance outside the home, like polishing to keep veneers at their whitest and shiniest. Mouths that have been treated for diseases like periodontitis can also benefit from more frequent periodontal maintenance appointments. The old "twice a year" usually doesn't apply to anyone and certainly not to people who have full mouth reconstruction. Be sure to talk with your dentist about the right plan to keep your smile healthy and beautiful for the long term.

If you experience any unexpected problems after your FMR, be sure to contact your dentist and let him or her know. You might need to schedule an appointment to make sure everything is working properly. FMR cases will often require multiple fine-tuning adjustment appointments.

DON'T – Forget to Wear a Prescribed Appliance

After your FMR is finished, your teeth may shift, which can cause problems with some of the work you've had done. This isn't necessarily because of the work itself – some people's teeth just tend to shift over time, while at other times, a patient's jaw needs time to adjust to the changes that were part of his or her FMR.

Regardless of *why* it happens, *when* it happens it can cause problems with your restorations, and may eventually force you back into your dentist's chair for replacements. For this reason, many dentists give their FMR patients retainers or night guards to anchor teeth securely in place and possibly protect against other types of damage. If your dentist gives you one, DON'T forget to wear it.

DO – Wear a Mouth Guard If You Play Sports

It may not look particularly attractive, however, a mouth guard can keep what's underneath (your teeth) safe from balls, elbows, hockey pucks and other tooth-damaging items. That will keep your FMR attractive, not to mention healthy and functional.

DON'T – Eat Between Meals

Your mother was right. Snacking between meals is unhealthy – especially when it comes to your teeth. Scientific studies actually show that people who nibble all day long constantly expose their teeth to bacteria, even if they brush and floss regularly. If you need to snack, try to snack only a few times a day, and remember to at least rinse your mouth with water afterwards.

In addition, it if you are one of those people that always needs to be sipping a beverage, this can be very damaging. Just about anything, other than water, will do damage to your teeth if you sip it constantly -- you are essentially bathing your teeth in your beverage of choice all day long. If the drink is soda, juice, coffee, or anything else acidic or sugary, it can be a major problem. If you have one of these drinks, finish it up and follow it with a glass of water. Drinking water is actually better than brushing when it comes to dealing with acidic foods and beverages – brushing immediately will increase the erosive effects of the acid.

DO – Chew Gum

Then again, maybe your mother was also wrong. You might be surprised to learn that chewing gum is good for your teeth. The reason is that when you chew gum, it stimulates your saliva, which is the natural way your mouth keeps your teeth clean. Just be sure the gum you choose to chew is sugar-free. Some types of gum even have an anti-cavity sugar substitute (Xylitol). A word of caution, though…if you have jaw problems such as TMD, chewing gum is usually NOT advised as it can worsen the problems you are having with your jaw joints.

DO Brush – But DON'T Brush Too Hard!

The most important thing you can do for your teeth is to brush them at least twice a day (and floss once a day). However, you can overdo it, especially if you have restorations like bonding, implants or veneers.

When choosing a toothbrush, always choose a soft-bristled brush. Essentially all dentists and dental associations recommend them, and I still find it strange that manufacturers even make other kinds of brushes. They aren't really good for natural teeth, and are especially hard on dental restorations. When in doubt, ask your hygienist or dentist what they would recommend.

The right toothpaste is also important. The goal is to avoid abrasion, which scratches and eventually dulls the porcelain used in some veneers and crowns. If you have veneers or porcelain crowns, this means skipping most toothpastes that identify themselves as "whitening," as they are usually highly abrasive. Again, if your dentist sends you home with a tube of toothpaste, the brand and formulation he or she chose is probably right for you and your new smile. If you have any questions, someone at your dentist's office should be able to guide you toward the right choice.

DON'T – Trust Your Professional Care to Just Any Dentist

You are likely to invest a significant amount of time and money in your Full Mouth reconstruction. So it only makes sense that you should protect that investment by trusting its care and maintenance to a qualified team. That means hygienists and staff who understand the type of work you have had done and know how to keep any restorations looking their best. The reason is that many of the routine dental procedures your family dentist and staff might perform can actually harm certain kinds of restorations. Before you open your mouth for just anyone, be sure that they understand exactly what is in your mouth, and how to clean and maintain it safely. For example, if you are at a high-tech office, your hygienist will likely use a Prophy-Jet® or an ultrasonic scaler. This could be a problem if they aren't completely knowledgeable in all aspects of your dental work. The strong blast from a spray cleaner can break the glazed surface of porcelain crowns or veneers, so that they eventually dull and stain more easily. Scalers can chip at the edges of crowns, veneers or bonding. Even the normal pumice polish most hygienists use during cleanings can scratch porcelain and composite. Your high-tech hygienist should have the knowledge to properly use all their equipment around your restorations and use special non-abrasive polish on your veneers and porcelain.

Fluoride treatments can also be an issue. So-called "Acidulated fluoride" can damage restorations while *neutral* fluoride will not. Neutral fluoride is not as strong, but it is clearly preferable for post FMR treatments as it will not stain your restorations.

Remember, you've likely invested a lot of your time and money in your Full Mouth Reconstruction, so it makes sense to do everything you can to take care of your new smile, both at home and professionally. Follow the tips in this chapter, and you should be enjoying your beautiful, healthy, functional new smile for many years to come.

13
A Few FMR Success Stories

As I mentioned when I started this book, to me, the best thing about performing Full Mouth Reconstruction is the fact that it really does change lives. It isn't just about making someone look better by giving him or her the kind of smile you might see on a movie or TV star. It's about actually improving people's health, improving their quality of life, in some cases eliminating pain, improving the way the jaw system functions, and even adding years to some people's lives. That's the kind of end result that only a comprehensive approach to dentistry – taking all factors into account and addressing both the causes and symptoms of problems – can achieve.

The feeling I get from helping to make that happen for my patients is just incredible. However, to really understand the changes in their lives, I'm going to let some of my actual full mouth reconstruction patients tell you their stories themselves.

Bev's Story

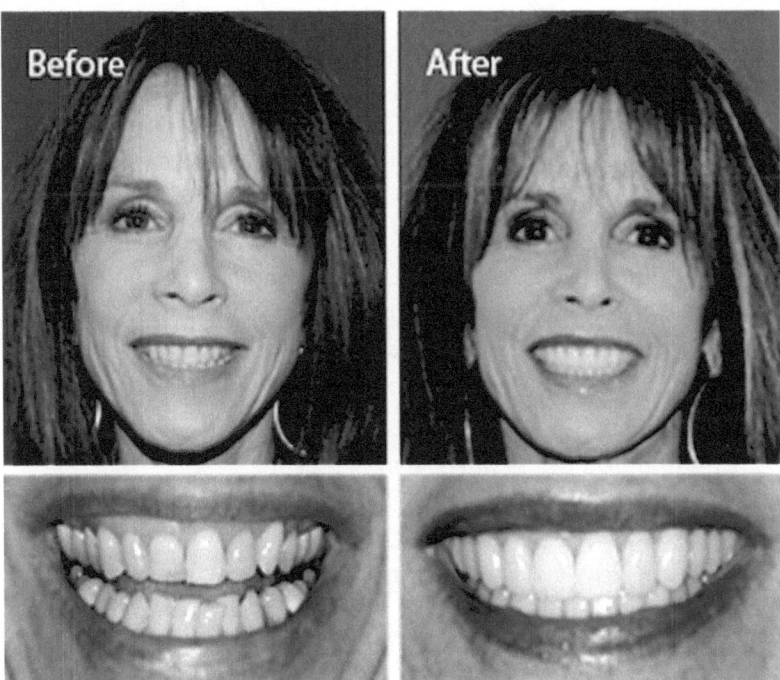

Bev came to me wanting a nicer smile. She had some jaw dysfunction issues along with cosmetic challenges, and had seen many different dentists over the years seeking their opinions.

I recommended that Bev consider orthodontics prior to veneers to get the best improvement both functionally and aesthetically — her upper and lower jaw arches are a little narrow, her teeth were not in ideal angulations and were crowded. She wanted other options than orthodontics. She chose to undergo a neuromuscular workup to help determine if there would be a better possible bite position, one that we could achieve WITHOUT doing orthodontics, that would also allow us to give her the kind of aesthetic improvement she desired.

I used a bite repositioning device for a few months to stabilize her in the best possible position we could achieve without orthodontics. Once this better functional position was established, I was able to provide her with her desired look as well. Her restorations included traditional veneers combined with porcelain crowns and partial crowns to achieve this better bite position, replace her old failing dentistry, and create her beautiful new smile.

My childhood experience of seeing the dentist was very negative -- filled with anxiety and pain. I distinctly remember the smell and sounds in Dr. Golden's office and the fear I felt as I read <u>Highlights Magazine</u> while waiting to be called in. Although they gave me Novocain, I recall always feeling extreme pain during the experience, constantly shifting in the chair to try to get away from the drill. Looking back, I know the dentist had to realize I was in pain, but he did nothing to make me more comfortable. This was especially difficult since I was prone to decay; I was at the dentist a lot as a child.

From my childhood on, dentists always told me I had a "bad mouth' – so I kept up with regular dental care despite my discomfort. My dentists left me with an understanding that my teeth were unhealthy and I had a lot of issues in my mouth. I didn't have orthodontic attention or good professional hygiene as a child, and my adult experiences seemed to continue that trend.

When I was in my thirties, I finally found a dentist who made me comfortable in the chair for the first time. He assured me if I felt any discomfort he would stop the procedure and give me more Novocain until I was pain free. He promised he would do everything to help me keep my teeth, and I was optimistic about the future of my smile for the first time. Unfortunately, that ended when the dentist was in a horrible accident and had to give up his practice. From then on, it became a downhill journey to find a dentist that gave me any sense of optimism or concern dealing with my many dental issues.

Those issues were numerous. I had a lot of decay, so there were many fillings – a lot of which were large and needed to be replaced with crowns. I had a bad chipped tooth in the front of my mouth that had to be bonded over and over. I also had many root canals, including one horrific experience where the sedation didn't block the pain and I wound up having emergency oral surgery on an abscess.

My teeth were also very crowded on the bottom, and as I got older my upper teeth shifted with more crowding and teeth angling in. They were also yellowed and did not respond to whitening because I had so many different surfaces with the crowns and bonding. As I headed into my forties and fifties, I became very self conscious of my smile and the appearance of my mouth.

By this time "perfect teeth" were the norm on the screen and in print. I began to search for a dentist who could help me improve my smile. I did everything from going to Dr. Dorfman from the TV show "Extreme Makeover" for cosmetic work to visiting my younger son's orthodontist to consider braces. At the same time, I was still trying to find a good local dentist to handle the basic care of my mouth.

One day, while waiting in a doctor's office, I was reading a local magazine called <u>Westlake</u> (named after the community we live in). I spotted an attractive full page advertisement for Westlake Dental Arts and noticed a very qualified female dentist with strong credentials in cosmetic dentistry. I went ahead and scheduled a consultation with Dr. Carson.

Upon meeting her, she listened to my history and carefully looked at my mouth. She spent a lot of time with me, discussing general care for my mouth and possibilities for restoration. I instantly liked her. She was warm, personable and very positive, and her staff was so friendly. She set out to give me detailed options for cosmetic work. I felt like she genuinely cared. I decided she was going to be my dentist on that first visit.

Dr. Carson and her staff immediately set out to get my mouth healthy. Working with her hygienist Carrie, I learned how to care for my gum tissue and I got REAL results. My gums became healthy and my daily dental hygiene regimen made a huge difference. At the same time, Dr. Carson gave me a device to reposition my bite. I was feeling positive about my mouth and the whole dental experience. By time I made the decision to have full restoration veneers, there was no question that Dr. Carson was going to do the work.

The process took months including time for Dr. Carson to have her baby. It was a gradual unveiling of my new smile. Yet even my temporaries looked fabulous. Seeing the progress with each session was so exciting, and seeing my finished new smile was amazing. I couldn't believe what an incredible job she had done. My teeth were straight and white. My mouth has never looked as healthy and beautiful as it does today.

Today, I am enjoying a new confidence in my smile that I never had. I enjoy being in pictures -- my mouth looks healthy and beautiful. My only regret with this whole process is that I didn't find Dr. Carson and her amazing staff sooner. I'm intensely grateful to Dr. Carson for the beautiful work she did and her dedicated philosophy.

Alan's Story

Alan really came to me for health reasons – his sleep apnea was actually threatening his life. Orthodontia was used to expand his arches and create the

proper space for his teeth. Oral surgery was necessary to correct his jaw joint disorder and to structurally correct him so that his airway improved. Once his orthodontia and oral surgery were complete, I approached his dentistry from a neuromuscular position to make sure that his teeth were in proper position to function best with his jaw position and muscles. Finally, we gave him a smile that looked as good as it functioned by replacing his old dentistry with crowns and veneers.

In late 2004, my managers at work told me -- in no uncertain terms -- that I was observed numerous times falling asleep in meetings, often when seated directly across from a client. I saw my internist, who referred me to a sleep specialist, Dr. Popper, who diagnosed me with severe obstructive sleep apnea. The only guaranteed treatment he could offer was a CPAP breathing machine, which I accepted and tolerated very well.

Even with the machine, however, I still tired easily and needed a midday nap. Then, a few weeks later, while lounging at home, I ate a large quantity of red licorice that had gotten slightly hard. All this heavy chewing caused some jaw pain so I visited Dr. Reider, who was my dentist at the time. I didn't realize it then, but that visit to the dentist probably saved my life.

We got on the subject of my sleep apnea and he referred me to Dr. Hang, an orthodontist, for help. Dr. Hang examined me and told me I had three choices. 1) I could continue to use the CPAP machine, and would still likely die 20 years younger than my father or grandfather (who both lasted into their '90s) due to my sleep apnea. 2) I could let him widen my upper and lower jaw to allow more room for my tongue to move forward, which would possibly buy me an extra 10 years; or (3) I could let him widen my jaws, then have the lower jaw advanced forward surgically, and then I'd have a chance to live a similar life span as my father.

I chose option 3.

Dr. Hang was able to cure my sleep apnea, however, following my orthodontia and surgery, my molars no longer meshed as they had prior to treatment. Dr. Hang told me that no one could do a better job restoring my chewing than Dr. Carson. After examining me,

Dr. Carson recommended a full mouth reconstruction, rather than just fixing the rear molars.

I had every tooth in my mouth, all 27 (having lost a molar as a teenager and having the wisdom teeth extracted as a graduate student) either capped with new crowns (replacing my old dentistry) or veneers placed over the existing teeth. All told, the planning and office visits took only a few months. The upper jaw was done first, in two visits, one to prep and another to place everything. Then, a few weeks later, the lowers were done. There were a few minor adjustments needed as I got used to the new teeth.

The general improvement in my overall health, especially not having sleep apnea, has been a godsend. No longer being tired all day, not falling asleep or needing naps is wonderful. The shape of my face – especially my profile – has been improved. That, plus always being complimented on my great smile and nice white teeth certainly helps my ego! I've never considered myself to be especially good looking, but with my improved health, my new smile and great looking teeth, I can't complain.

Maria's Story

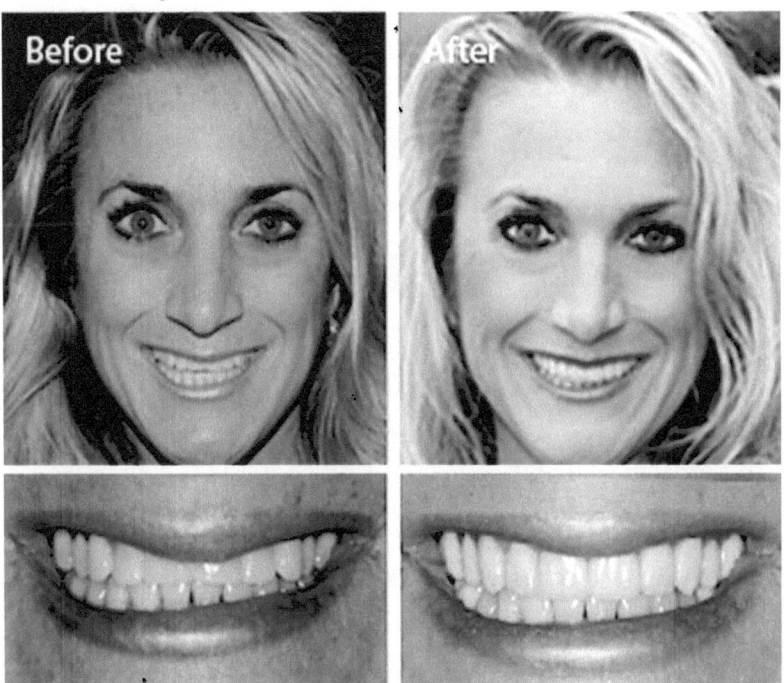

First of all, full disclosure here -- Maria is my assistant. She wanted to address bonding that was continually chipping and wear on her teeth that was getting worse. She also had neck pain and clicking and popping of her TMJ's, but didn't suspect that this had anything to do with her bite.

At first, she was only interested in a few front veneers for the chipping, then she decided she didn't like how the sides of her smile caved in (she had premolars extracted for childhood orthodontic treatment that often results in the narrowing of a smile). She thought she'd like 10 upper veneers for cosmetic purposes only, but agreed to let me do a neuromuscular workup to evaluate her bite first.

Once I determined a better functional position for her bite, I placed her in a bite repositioning orthotic to see how she did. Her neck pain went away, she could turn her head further in both directions, and the clicking went away - she was immediately more comfortable. We were able to restore her to this new bite position AND give her a more beautiful smile by doing crowns on her upper and lower molar teeth and traditional veneers on the rest of her upper teeth. We bleached her lower teeth to give her the smile she has today.

I've been in dentistry 30 years, so I know the pros and cons of dental treatment, why it's important to go to the dentist, and why you need to take good care of your teeth.

My issue was pretty simple – I thought. My brother had popped me in the front tooth as a kid and I had a little chip in my two front teeth for years. Dr. Carson had bonded it 8 years ago, but with time, the bonding wasn't holding up as well as it used to. In addition, I realized my front teeth were getting shorter and shorter and thinner and thinner. We started discussing putting some veneers on them.

At the same time, I had some pretty worn teeth in the back – although I didn't have any discomfort that I was actually aware of and I didn't really realize they were as worn as they were. I also had a lot of headaches – Jenny – that's our other assistant – and I were always reaching for the Advil. It was part of the daily routine. I have teenagers, so I figured it was just a part of life to have headaches and neck aches and stress and sleepless nights. I figured it was my kids, not my teeth– even though I work in dentistry!

Dr. Carson did a neuromuscular workup on me and you could see that my muscles were very tense and my bite was off. This meant that if Dr. Carson put some pretty veneers on my teeth, there was a good chance that my function would just destroy them!

When we took impressions of my teeth and poured my models, I got to see how much anatomy I had literally just chewed through, and that my nose and my chin had started getting closer and closer together. I was so afraid I was going to have that "witch look".

As a result, what we were thinking might be two veneers to fix the chip, then ten veneers to broaden my smile, turned into Dr. Carson rebuilding my bite. She put some crowns on my back teeth and reopened my bite. She also put me in orthotics for months to get my bite in position and stabilized.

Then it was on to the veneers.

I always liked me smile, but I kind of had that "social six" look where my front six teeth looked great and then it dipped in from having my bicuspids pulled when I had braces as a kid. I wanted to "bulk it out" so you could see a broader smile.

We decided to do 12 veneers across my upper teeth. I was a nightmare patient, knowing too much about dentistry and trying a few different smile designs for me. I tried a flatter, straighter look; I tried a rounder look; I was in three different sets of temporary smile designs. I did have a "wow moment" when I saw a photo of myself in one of the sets of temporaries at a Halloween party, and I LOVED my smile. I looked at the photo and I said, "Damn! I have a great smile!" I was just in my temporaries! I told Dr. Carson, "This is exactly what I want." She gave my teeth some length, broadened my smile, kept the characteristic marks that I liked, and now I have my new smile.

It's funny, I always liked my smile, but I flash it a lot more now. Except for the people in the office, very few know my secret. Even my parents have no idea that I had this dental work done. Every one just says that I look great. That's the sign of a job well done!

The funniest part is, after Dr. Carson put me in this new bite position, I didn't reach for the Advil anymore. We see it in our patients all the time, the headaches and the neck pain, but I didn't categorize myself as a "pain patient," ever! But once she put me in the correct bite, I have had no headaches. It's made a huge difference.

Today, when I see patients, I can say, I've gone through it. This is what you're going to feel, this is what it's going to look like, and I've been there. I think that's huge, and I think it comes across in the way we discuss dentistry today.

Joan's Story

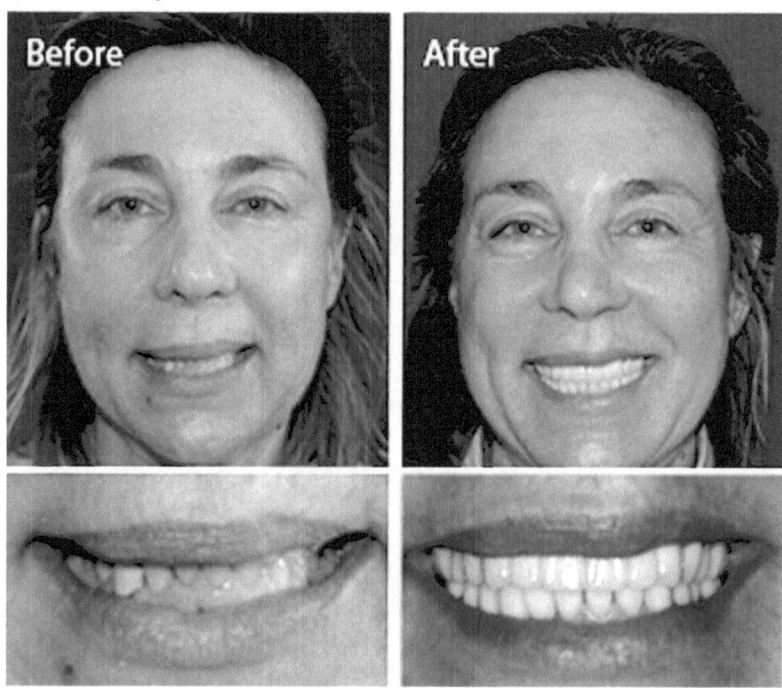

Twenty years before coming to see me, Joan suffered a serious, on-the-job accident. She was (and is) a veterinarian, and was kicked in the mouth by a horse – an incident that knocked her unconscious and resulted in injuries ranging from brain trauma and a broken nose to losing several teeth. Dentists had done what they could to restore her smile over the years, but her options had been limited because of the damage that had been done.

We set out to give Joan a beautiful, functional smile using implants, traditional veneers, and replacement dentistry of porcelain crowns and partial crowns, all focused on rebuilding her to a better bite from a neuromuscular perspective. We worked with a physical therapist while she was in the bite repositioning appliance to get the best possible results.

After my accident, I focused more on my facial scars and brain injury and felt my teeth were hidden scars. I was glad to have any teeth at all. I so appreciated just having teeth, I didn't realize my self-esteem was suffering. My veterinary practice doesn't allow for a lot of "luxuries," and I thought a better-looking smile would be just that – a luxury. I now know that having teeth, and good teeth, is not a luxury. They are a necessity.

Before I met Dr. Carson, the teeth I had and a bridge that I thought had been very well done were causing problems. My bite was not in balance as my bridge placed pressure on my other teeth, which led to TMJ discomfort and weekly migraines. As far as appearance went, my teeth were also discolored, and those colors varied. I went to Dr. Carson hoping to change that and finally to have a beautiful smile.

I needed bone grafts to support the implants, so the process took a while. After the restorations were completed, I loved my teeth, but needed to wait to save more money to have the remaining teeth the same color. When all was completed, I LOVED MY MOUTH!!!!!!

Dr. Carson went far beyond just giving me a beautiful smile. She recommended physical therapy to work with the neuromuscular dentistry to correct the TMJ problem and help with my bite position, which stopped most of my migraines.

My accident took place when I was 35 years old. Now, I have lovely teeth thanks to Dr. Carson – ever since I was 55 years young. Today, I still work on horses -- I do mobile veterinary work, and often refer my small animal patients to clinics when needed. Knowing what I now know, I have the confidence to look a client in the eye, smile and recommend a procedure because it is medically necessary. My clients smile back -- and usually, they comment on my beautiful teeth!

George's Story

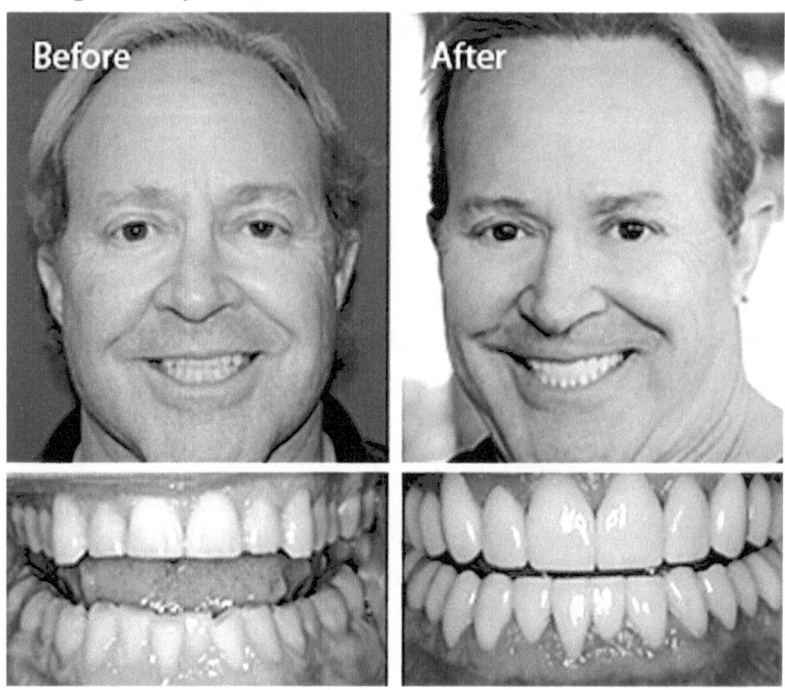

George is a body builder who came to see me for a performance mouth guard called a PPM, which is based on neuromuscular principles of a better bite and leads to improved balance, strength, and flexibility. He was very happy with the results of the PPM, and then expressed some interest in improving his smile.

At first he was just thinking about his upper front teeth. We showed him that when he talked and smiled he actually showed mostly lower teeth. We then discussed how the PPM is based on neuromuscular principles, and that we could use those same principals to improve his bite permanently.

George realized that improving his bite while getting the best looking result was the best way to go. We used neuromuscular principles to rebuild the position of his bite, placed minimal prep veneers where possible (on most of his teeth), and used porcelain onlay restorations on his molar teeth.

I had been avoiding the dentist for about four years -- primarily out of procrastination, yet also because my last dentist did some work that had to be redone. So I was basically disenchanted with dentists in general.

I had not experienced any problems that I knew about when I went to see Dr. Carson. I went to her because I am an athlete and was interested in what is called a PPM – that stands for Pure Power Mouthguard – that aligns/realigns your jaw so that nerve transmission down the spinal cord is optimized, the airway is more open, and it all translates to greater athletic performance.

At the appointment, Dr. Carson pointed out that even though I was not experiencing any problems at the moment, in the future I could be faced with multiple root canals due to the grinding and clenching of my teeth over the years and the consequential loss of enamel and fracturing.

Dr. Carson repositioned my bite and placed veneers on all my teeth. When I finally saw my new smile, it was a "monster moment" for me. I was floored!!!! I hadn't considered my smile as that "lacking" (for lack of a better word). My teeth weren't that "dingy" in color (or so I thought!). But, when she was finished and she held up a mirror.....well let me say this....if wasn't so macho I would have cried for joy!

To be truthful, I've had other cosmetic surgeries in the past. And I was very happy with their outcomes of helping to keep my appearance as youthful and pleasing as possible. Yet, when it had come to my smile, I didn't think it was worth the money. I know a beautiful smile is important. Yet, still I felt the cost/benefit ratio was not there.

I couldn't have been more WRONG! All my friends and family have commented about how wonderful and striking my smile is! And I get to cure, and save the cost of, a massive future problem.

Today, I smile more. A lot more. Actually, all the time. I'm proud of how my smile looks and certainly don't mind sharing the look of it with others. Now don't misconstrue me - my newfound confidence in my appearance is not a cocky arrogance, yet merely a quiet self-assuredness that I "look my best". And that's all anyone could ever ask for.

Susan's Story

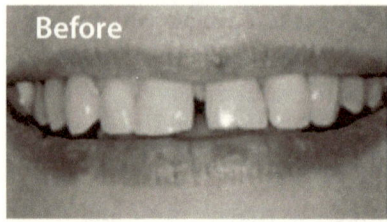 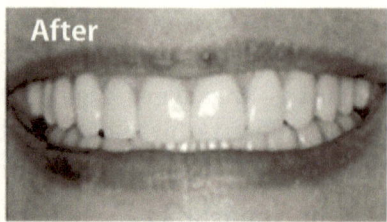

Susan's case was very serious – her health was severely compromised by her bite and the position of her jaws. Restoring her smile, and health, involved orthodontia, oral surgery, and finally traditional veneers on her front teeth and replacement dentistry for her old dentistry, approached from a neuromuscular standpoint. We worked with a physical therapist and a massage therapist during the orthotic phase (bite repositioning) to achieve the best possible results.

I had always taken care of my teeth. So I was shocked when Dr. Carson looked at me with concern and told me that if we didn't correct my bite, I would be looking at implants or dentures in as soon as five years. It was a complete surprise to learn that for years, I had been clenching and grinding my teeth -- and that this was causing major structural problems.

I also had a diastoma (space) between my two front teeth that seemed to be getting larger, and gaps were developing between my incisors and front teeth. My front teeth were also chipping on the bottom edges.

Although I'm pretty sure she knew what Dr. Hang was going to recommend, all she mentioned to me was an orthodontic consult -- which seemed innocuous. But Dr. Hang's diagnosis threw me for a loop.

Dr. Carson helped me piece the picture together – and it was a frightening portrait of me. I need braces to reposition my teeth, but braces couldn't repair the years of wear and tear. Crowns and veneers could restore my teeth to a healthier condition, but would do nothing to address my current, sorry state.

Although I looked "normal," my jaws were positioned in a way that narrowed my airway to less than ¼ of normal size. This led to sleep apnea, constant tiredness, weight gain and pre-diabetes – in addition to the problems with my teeth.

Without a doubt, without these two very special professionals in my life, I would now weigh more than 200 pounds, inject daily insulin, sleep my weekends away and, oh yes…drop my teeth in a glass as I crawled into bed.

Instead, I had braces, followed by major surgery -- there were 12-15 different procedures done on my face, jaw and soft tissues. After this, Dr. Carson did a full mouth restoration on me. All my teeth are either veneers or porcelain crowns. Once I decided to have the surgery (this took more than a year), I think everything was wrapped up in about 2 years.

When my restorations were done, I was shocked by my appearance – My teeth were beautiful, but I no longer looked like "me." My reasons for proceeding with the restoration were 100% about health -- not appearances. However, it didn't take long before I was in love with my new smile.

The surgery, not the dental work, changed my life -- I no longer sleep with a CPAP machine. But really for me, the orthodontics, surgery, and dental work were all of one piece. They all contributed to an end result that could not have been achieved if one of the pieces had been eliminated.

Not every Full Mouth Reconstruction experience is quite as dramatic as what these six patients went through. Yours may be a simple procedure that improves the look and function of your teeth, or it may actually save your life. What's important to remember is that with FMR, your treatment plan will be custom-created around you and for you, using the most modern, state-of-the-art thinking and technology to give you the best possible smile – and maybe even the best possible life.

After all, you deserve nothing less.